Narratives on Teaching and Teacher Education

Narratives on Teaching and Teacher Education

An International Perspective

Edited by
Andrea M.A. Mattos

palgrave
macmillan

NARRATIVES ON TEACHING AND TEACHER EDUCATION
Copyright © Andrea M.A. Mattos, 2009.

All rights reserved.

First published in 2009 by
PALGRAVE MACMILLAN®
in the United States—a division of St. Martin's Press LLC,
175 Fifth Avenue, New York, NY 10010.

Where this book is distributed in the UK, Europe and the rest of the world, this is by Palgrave Macmillan, a division of Macmillan Publishers Limited, registered in England, company number 785998, of Houndmills, Basingstoke, Hampshire RG21 6XS.

Palgrave Macmillan is the global academic imprint of the above companies and has companies and representatives throughout the world.

Palgrave® and Macmillan® are registered trademarks in the United States, the United Kingdom, Europe and other countries.

ISBN: 978–0–230–61233–4

Library of Congress Cataloging-in-Publication Data

    Narratives on teaching and teacher education : an international perspective / edited by Andrea M.A. Mattos.
      p. cm.
    Includes bibliographical references and index.
    ISBN 0–230–61233–4
      1. Teaching—Cross-cultural studies. 2. Teachers—Training of—Cross-cultural studies. I. Mattos, Andrea M. A.

LB1025.3.N37 2009
371.102—dc22   2008051593

A catalogue record of the book is available from the British Library.

Design by Newgen Imaging Systems (P) Ltd., Chennai, India.

First edition: July 2009

10 9 8 7 6 5 4 3 2 1

Printed in the United States of America.

*To my husband,
the one person who has taught me to look
at life from a positive stand point
and to never give up.*

Contents

List of Figures and Tables	ix
Foreword: Inquiring into Narrative Matters D. Jean Clandinin	xi
Acknowledgments	xv
Introduction Andrea M.A. Mattos	1
1 Narrative Frameworks for Living, Learning, Researching, and Teaching Anne Laura Forsythe Moore	11

Part I Stories of Discovery and Transformation

2 Understanding Classroom Experiences: Listening to Stories in order to Tell Stories Andrea M.A. Mattos	31
3 Becoming a Teacher: Using Narratives to Develop a Professional Stance of Teaching Science Robert W. Blake Jr. and Sarah Haines	47
4 Personal Politics and Identity in Student Teachers' Stories of Learning to Teach Alan Ovens	65
5 Learning to Teach Across Cultural Boundaries Neil Hooley and Maureen Ryan	77

6 River Journeys: Narrative Accounts of South Australian
  Preservice Teachers during Professional Experience   91
  Faye McCallum and Brenton Prosser

7 Exploring Ways of Promoting an Equality Discourse
  Using Non-Text/Creative Approaches for Learning in
  the Everyday Lives of Adult Literacy Learners   107
  Rob Mark

8 Team Teaching: Having the Eyes to See the Wind   123
  Dawn Garbett and Rena Heap

9 Enhancing Faculty Commitment, Hope, and
  Renewal through Developmental Performance Review   137
  Georgia Quartaro and Bob Cox

Part II Stories of Hope

10 "It Gives Me a Kind of Grounding": Two University
   Educators' Narratives of Hope in Worklife   151
   Denise J. Larsen

11 Finding the Time and Space to Write:
   Some Stories from Canadian Teacher Educators   167
   Dianne M. Miller

12 Learning about Hope through Hope:
   Reflections on the ESL Enterprise   177
   Judy Sillito

13 A Newcomer's Hope: A Narrative Inquiry into One Teacher
   Educator's Professional Development Experiences in Canada 191
   Yi Li

14 A Tail of Hope: Preservice Teachers' Stories of
   Expectation Toward the Profession   203
   Andrea M.A. Mattos

Notes on Contributors   217
Index   221

List of Figures and Tables

Figures

1.1	The Hingley Homestead, Caribou, Pictou County, Nova Scotia, Canada	18
2.1	Elements that Influence the FL Classroom	42
6.1	Example River Journey	98

Tables

2.1	Subcategories for the Participant's Dominant Stories	37
3.1	Source of Written Narratives	51
3.2	Comparing Enduring Beliefs	56
3.3	Jennifer's Qualities of a "Good" Teacher	60

FOREWORD

Inquiring into Narrative Matters

D. Jean Clandinin

The chapters in this lovely edited collection call me to consider the importance of the multiplicity of ways of engaging in narrative inquiry. Working from the following understandings of narrative inquiry, I see the range of ways the authors whose work is represented in this collection have taken up their inquiries. All authors appear to work from "a view of human experience in which humans, individually and socially, lead storied lives" (Connelly and Clandinin 2006, p. 7). Narrative inquiry is a way of understanding and inquiring into experience through "collaboration between researcher and participants, over time, in a place or series of places, and in social interaction with milieus" (Clandinin and Connelly 2000, p. 20).

Three commonplaces of narrative inquiry, temporality, sociality, and place, specify dimensions of an inquiry and serve as a conceptual framework for narrative inquiry. In our understanding (Connelly and Clandinin 2006) of these three commonplaces, they are places or dimensions that need to be simultaneously explored in undertaking a narrative inquiry. Inquiring into experience through attending to all three commonplaces simultaneously is, in part, what distinguishes narrative inquiry from other methodologies. Through attending to the commonplaces, narrative inquirers are able to study the complexity of the relational composition of people's lived experiences both inside and outside of an inquiry and, as well, to imagine the future possibilities of these lives.

The editor of this collection, Andrea Mattos, has thoughtfully organized the book into two sections: stories of discovery and transformation and stories of hope. What strikes me about this organization revolves around how temporality is foregrounded in the study of experience as she positioned these chapters in relation with one another within the book.

Temporality is a key dimension of understanding experience narratively and is, as noted above, central to any narrative inquiry. As one inquires into the lived and told stories of teachers and teacher educators, temporality, that is, the past, present, and future, is a dimension that needs to be always visible in the inquiry. As I noted above, however, inquiry also needs to attend simultaneously to the commonplaces of sociality and place. All of the chapter authors work with close attention to the simultaneous exploration of the three commonplaces. However, temporality becomes foregrounded by the way Mattos organizes the book. She foregrounds temporality by drawing attention to hope as a narrative composition and, in so doing, gives a sense of an intentional trajectory of storied plot lines. By focusing on hope in the second part of the book, Mattos makes explicit the focus on the future within the temporal dimension of narrative inquiry.

In Part I, we see the power and importance of engaging teachers and teacher educators in inquiring into their past and present stories. In some chapters we see researchers engaged in analysis of student teachers' and teachers' narratives of experience. In other chapters we see researchers engaging in more autobiographical narrative inquiries as they inquire into their own narratives of experience as teacher educators. The chapters in Part I are rich in acknowledging the sociality dimension with its dialectic between the personal and the social. Similarly the chapters attend to the cultural, institutional, and linguistic narratives that shape the landscapes on which the individual teachers, teacher educators, and researchers live.

What marks the second part of the collection is a different attention to temporality, a temporality that attends to a future-oriented trajectory of hope. In this part of the book, the chapters also attend to the dimension of sociality with its dialectic between the personal and social shaped by institutional, cultural, social, and linguistic narratives. The chapters here, however, while grounded in stories of the past and present experiences of teachers, teacher educators, participants, and researchers, make more explicit the future temporal dimensions of narrative inquiries. Understanding hope narratively allows these researchers to point out possible future plotlines, as Mattos notes "a sense of new possibilities and expectations that enhance their empowerment," which point toward, as Maxine Greene (2008) suggests, "what I am not yet." There is a sense of becoming; becoming that follows an upward trajectory. The chapters, too, help us see hopelessness, a trajectory that points downward into the future with a sense of loss, of not becoming.

Mattos thoughtfully adds to the richness of the chapters by her conceptual organization. The chapters in and of themselves also offer readers much to consider. One way to read them is for their theoretical insights into narrative conceptualizations of teaching, of teacher education, and of hope. Another way to read them is as sources of what Eva Hoffman (1994) calls resonant remembering. A third way of reading them is as models of possibility, that is, as ways we might engage our own teacher education practices, teaching, or research projects. Whatever our intentions and purposes are for reading this book, the various chapters written by a diverse international audience of researchers offer us a multiplicity of possibilities for entering conversations around narrative inquiry.

References

Clandinin, D.J., and F.M. Connelly. 2000. *Narrative inquiry: Experience and story in qualitative research*. San Francisco: Jossey-Bass Publishers.

Connelly, F.M., and D.J. Clandinin. 2006. Narrative inquiry. In J. Green, G. Camilli, and P. Elmore (Eds.). *Handbook of complementary methods in education research* (pp. 375–85). Mahwah, NJ: Lawrence Erlbaum.

Greene, M. 2008. *From bare facts to intellectual possibility: The leap of imagination: A conversation with Maxine Greene*. Paper presented at the American Educational Research Association Annual Meeting, New York.

Hoffman, E. 1994. Let memory speak. *The New York Times Book Review*, January 23.

Acknowledgments

I would like to thank all those who have contributed to the success and fulfillment of this project. First and foremost, I would like to thank Jill Lake, former editor at Palgrave Macmillan, recently retired, for being the first person to believe in this project. I would also like to thank Julia Cohen, my assistant editor at Palgrave Macmillan, for her enthusiasm and dedication, and for her careful support.

I believe no author or editor can accomplish such a big task as writing or editing a book without the help of innumerable people whose names are rarely mentioned or recognized. Unfortunately, once again it will not be possible to name all these people for reasons of time and space, but I want to acknowledge their effort and diligence in making this project possible and real. Finally, I would like to thank the contributors for their patience during the editing process and for their willingness to contribute to the book. Without their keen interest in the covered areas and commitment in meeting deadlines and complying with norms and instructions, it would not have been possible to compile this collection.

Introduction

Andrea M.A. Mattos

Indeed, it has been argued that the nature of reality itself as experienced by humans and human cultures is an emergent effect of narrative interactions.

—Jerome Bruner

Every culture has its main stories or "master narratives," as Allan Luke (2004) has mentioned. In my country, one such main story says that "every man must plant a tree, have a child, and write a book" in order to be considered a real man. In times when women reclaim positions and legitimacy in various—if not in all—sectors of modern life, I will allow myself to understand this main story not in terms of manhood but in terms of mankind. As a woman—at the same time product and producer of the culture in which I am inserted—I have cherished the possibility of some day being the main character in this story. Although I would not claim to be a green activist, I have planted some trees in my quest to contribute to the quality of the air we all breath. I have also had my own children, three young ladies, who fill my life with all the joy and laughter a mother could ever dream of. What was missing was the book.

The idea came when I first went to Canada, in 2006, to take part in the Narrative Matters Conference, in Wolfville, Nova Scotia. There, effective technology and impeccable organization were the background to the top-quality papers presented by specialists on a variety of areas. The wide international representation of participants was another highlight. As a researcher interested in the field of teacher education, I made a point of attending most sessions in the field. On the last day of the conference, the organizing committee announced that there would be no proceedings that year. I came back home with the disappointing feeling that researchers who had not been present at the conference would never have the chance to learn about the many papers presented. During the long-haul flight between Canada and Brazil, I thought of inviting the many researchers I had met to publish an edited collection on teacher education.

When I arrived in Brazil, I started working on this idea. Some of the invited researchers readily agreed to participate. Others kindly declined the invitation. I made further contacts and invited people who had not attended Narrative Matters, but who were interested in narrative research and teacher education. I sent out a call for chapters and waited. By the end of a couple of months I had collected more than twenty proposals for chapters from researchers in the five continents. I was stroke by the variety and coverage of the proposals and thought I had enough to compile a collection. Unfortunately, for several reasons not all the proposals could be included.

This is the story of this book, but it is not the end of the story yet. The remainder of this introduction will be dedicated to present and discuss narrative research and the various chapters in the collection. Although the several chapters do not always follow the same research background for collecting and analyzing data, they all adhere to some kind of narrative framework, which is the main bond that unites the chapters in this collection. The rest of the story comes next.

In recent years, research in various fields has assumed a more qualitative perspective. For this reason, there has been increasing interest in narrative modes of thought and expression. The so-called Narrative Turn has influenced the humanities, the social sciences, and the health sciences, and research along this methodological line has covered a wide variety of topics, from narrative in fiction to narrative modes of understanding human experience. This interest builds on the work of important theorists: the American psychologist and philosopher William James (1842–1910) has referred to two modes of thinking, the paradigmatic and the narrative mode. More recently, the cognitive psychologist Jerome Bruner (1986) referred to "narrative ways of knowing" and described narratives as stories, dramas, and historical accounts that emphasize human intention and action. Bruner's ideas have greatly influenced current research in the fields of psychology, sociology, education, linguistics, and applied linguistics, to mention but a few.

He states that humans have the "capacity to organize and communicate experience in a narrative form" (Bruner 2002, p. 16). In the author's point of view, it is in constructing stories and myths, and in listening to the stories of others, that we deal with our experience and make sense of our reality. According to Bruner (1990, 2002), narratives constitute and transform us into who we are, that is, it is through telling and listening to stories, including our own, that we are continually formed and transformed.

As Bruner (2002, p. 25) asserts, "the sharing of common stories creates an interpretive community [that promotes] cultural cohesion." He defines culture as "shared symbolic systems" and "traditionalized ways of living and working together" (Bruner 1990, p. 11). The author states that culture is "the major factor in giving form to the minds of those living under its sway" (p. 12) and it is culture and the search for meaning that mould and guide the changing nature of the human species. Besides, it is through narratives that "the human being achieves (or realizes) the ability not only to mark what is culturally canonical but to account for deviations that can be incorporated in narrative" (p. 68).

In their very influential book *Narrative inquiry: Experience and story in qualitative research*, Clandinin and Connelly (2000) build on the work of John Dewey and other theorists to construct and demonstrate a research framework widely used among social scientists interested in conducting narrative research. Their ideas have impacted many modern researchers, especially in the field of education, as they believe, with Dewey, that "examining experience is the key to education" (p. xiii).

This collection is in line with this recent perspective. The overarching theme that brings together all the chapters in this collection is teaching and teacher education. *Narratives on teaching and teacher education: An international perspective* is an edited collection of papers on lived experiences and stories of teaching and/or learning to teach, which focus on the subjective meanings (including affective and psychological matters, such as identities, attitudes, beliefs, and hopes) and interpretations evoked in teaching and learning to teach. Framed as it is at the intersection of narrative research and teacher education paradigms, *Narratives on teaching and teacher education: An international perspective* investigates the complex dynamics of teaching and learning to teach through multiple perspectives and theoretical frameworks. In providing space for discussion and reflection on the complexities, ambiguities, and possibilities of teaching and learning to teach, the collection contributes to increasing the depth and expansion of knowledge construction and meaning production in narrating teaching and teacher education. The papers compiled in *Narratives on teaching and teacher education: An international perspective* attempt to describe and shed light on the difficult and often contradictory routines of teaching and learning to teach, as they are captured in the narratives and stories that abound in everyday classrooms and educational settings.

The collection is divided in two parts. Part I is named Stories of Discovery and Transformation, and Part II, Stories of Hope. Chapter 1 was not included in any of these two parts because it offers an overarching context for narrative research in teaching and teacher education in general. In this chapter, Anne Laura Forsythe Moore explores teacher education and the teacher/learner relationship in terms of a *puzzle framed*. Drawing on the work of both John Dewey (1938) and Clandinin and Connelly (2000), she describes a narrative exercise that "encourages the practice of viewing our world with plural perspectives in a multi-dimensional way." Based on her experience as a teacher educator in Canada, Forsythe Moore shows how the phenomenon of awakening to a renewed narrative perspective of our own social history during the retelling of our personal and professional stories transformed as *reconstructed narratives* can help to illuminate the "meaningful connections between living, learning, researching and teaching." In exploring the power of literacy, language, and relationship in community, she asserts that an active engagement in reading, writing, and storytelling activities, and listening to stories, including our own, supports a personal and professional awareness that sustains a curriculum design that is democratic, equitable, and socially just. Forsythe Moore concludes by reaffirming the *power of reflexivity* in contributing to the teacher/learner relationship and how important this understanding is to the field of teacher knowledge, teacher education, and curriculum development.

Part I, as its name indicates, compiles chapters that describe the rich process of discovery and transformation generally experienced by preservice and in-service teachers whenever they decide to look at their everyday practices in classrooms and educational contexts with an inquisitive eye. Chapter 2, the first chapter in this part, written by Andrea M.A. Mattos, argues for a narrative understanding of the complexities of teaching English as a foreign language (EFL), a much debated area of research, especially in Brazil where the study was conducted. Through a number of reflective interviews that provide the space for teacher stories, she discusses the complex nature of the foreign language classroom, often disturbed by both *internal* and *external* influences, and reveals "a set of different elements that may influence the classroom context, interfering with the teacher's performance." Mattos concludes that teachers are often faced with problems and other sources of anxiety in their experiences, which may lead to "burnout" and other forms of teacher disease, and argues that narratives and stories are an important tool in understanding such problems.

In chapter 3, Robert W. Blake Jr. and Sarah Haines use written narratives as a means for preservice teachers to modify their personal theories of teaching science into a *Professional Stance*—a set of beliefs that novice teachers construct as they revise their personal theories in face of the realities of daily teaching in the United States. Blake and Haines argue that the strength of narratives is in the analysis and discussion of each narrator's own solutions and how they address and resolve their own conflicts. They believe that "this personalization of the teaching process will provide interns an opportunity to more critically analyze their own beliefs and views of what it means to become a teacher."

Alan Ovens, in chapter 4, also uses stories of student teachers but, unlike Blake and Haines' in the previous chapter, these stories are based on memories of prior experience. The author employs *memory-work* to examine the ongoing identity narratives of two student teachers in the context of a New Zealand teacher education program in physical education. Ovens investigates their written narratives and discusses how these narratives may represent formations of identity. He is also interested in considering how their "principled positions represent political subjectivities that mediate each student teacher's lived experience of learning to teach." The study reveals that it is possible to better understand how each student constructs the nature of teaching and the decisions they make concerning curriculum and pedagogy.

Chapter 5, by Neil Hooley and Maureen Ryan, is somewhat different from all the previous chapters. Instead of studying preservice teachers' narratives, the authors choose to narrate their own process of discovery and transformation as they struggled to "link major issues in the literature regarding Indigenous ways of knowing with the ideas and concepts of privileged non-Indigenous knowledge in the traditional curriculum" in Australia. Embracing the notion of *two-way learning* and drawing on the ideas of Dewey, Bruner, and Clandinin and Connelly, they describe an experience in establishing a Bachelor of Education program for Indigenous and non-Indigenous teachers, a unique opportunity in Australian education. Hooley and Ryan manage to design a narrative curriculum that functions as the backbone for what they call *two-way inquiry learning*. The chapter is a personalized account of the authors' attempt to "recognize and respect the knowledge and cultural practices of Indigenous communities," aiming at cross-cultural understanding and tolerance of more *Indigenous ways of knowing*.

Chapter 6 also refers to the Australian educational context but goes back to stories of preservice teachers. Faye McCallum and Brenton Prosser argue that preservice teachers in Australia usually begin their professional preparation "believing that what they need most is to expand their knowledge of curriculum content and to develop effective behavior management strategies." They soon discover, though, that teaching is much more complex and demanding than they had imagined. The authors report on an initiative with graduate entry students in Australia, using metaphor and narrative to explore professional experiences. They employ the metaphor of *river journeys* to provide the basis for reflection on "the aesthetic, affective, and storied elements of teacher identity" and to understand the "multiplicity and complexity of teacher identity development." In the conclusion, the authors insightfully reflect on the limitations of the *river journey* metaphor and suggest possibilities for improvement.

In chapter 7, Rob Mark examines how creative/non-text methodologies can be used to promote literacy learning among adults. Grounded on theories of literacy that focus on power relations and inequalities, the chapter describes an European funded project that encourages learners and tutors to explore such issues using non-text methods of learning. The project employed a number of innovative text-free methods to promote inclusion in learning through the exploration of equality issues affecting learners' lives in Ireland. Inspired by the postconflict situation in Northern Ireland, the project also aimed at contributing to peace building and reconciliation, through the exploration of inequalities and experiences of conflict focusing on individual stories from the lives of both tutors and learners. The author argues that as a result tutors and learners involved in the project were able to improve their understanding of "equality issues affecting their lives" and at the same time were able to develop their own knowledge and skills. This has a powerful message for the development of programs of learning in adult literacy education.

Chapter 8, by Dawn Garbett and Rena Heap, explores the interesting but challenging experience of team-teaching. The context of the experience is again the teaching of science, but here the authors draw on their own stories as teacher educators in a Graduate Diploma of Teaching course in New Zealand. During the course, they focused not only on content knowledge but also on the more subtle "subtext of learning about teaching." Dawn and Rena acknowledge the fact that they have found team-teaching very demanding both from personal and professional perspectives, but recognize that they have grown as teacher educators as a consequence of the professional conversations they maintained during the course. In this chapter, they share their findings from this transformative experience and offer suggestions for further successful team-teaching practices in the future.

In chapter 9, the last chapter in Part I, Georgia Quartaro and Bob Cox describe a new *strengths-based* faculty review process piloted in a community college in Canada, which centers on *excellence* in teaching and learning. The review process, involving more than twenty professors, attempted to be comprehensive, practical, and to provide an arena for self-reflection, using a range of activities such as peer-observation, summary meetings, and personal portfolios. The authors report on the results of this piloting experience, which encouraged self-reflection about teaching based on feedback from different sources, in order to "create an opportunity for engagement, reflection, and renewal." Quartaro and Cox provide a narrative analysis of the review process that considers both the individuals' stories and the interplay among the participants' accounts. Important themes in the area of teaching, such as renewal, optimism, hope, and, commitment are discussed and aspects of teaching that are rarely spoken about are presented and analyzed. The authors also address the drawbacks and limitations of the process but contend that it was generally positive and rewarding for all involved.

Part Two compiles narratives and stories of hope. The chapters in this part are mostly inspired by the work of researchers from the Hope Foundation of Alberta, in Canada, a research center affiliated with the University of Alberta, and a registered nonprofit organization dedicated to the study and enhancement of hope. In chapter 10, Denise J. Larsen, a leading researcher at the Hope Foundation, discusses the *deeply held hopes* about what being a teacher might mean. Through multiple interviews with university professors from Faculties of Education within Canada, and employing a narrative inquiry approach (Clandinin and Connelly 2000), she seeks to learn how meaningful worklife hope is lived within the university. Larsen argues that "teaching contexts do not always support educators' heartfelt worklife quests" as competition, unhealthy relationships, and personal unhappiness tend to dominate educational institutions. In her data, Larsen identifies key narratives related to hope, including deeply held stories of purpose, as well as multiple narratives that had the power to either sustain or threaten hope in academic settings.

Chapter 11, by Dianne M. Miller, draws on the results of a wider study on Canadian teacher educators' experiences of changing institutional and professional demands, including the emergence of a research culture. Miller focuses especially on the issue of "finding the time and space to write," an aspect of the complex negotiations that individual teacher educators undertake as they strive to fulfill job requirements in the changing university context. Closely studying more than twenty interviews, Miller discusses the teacher educators' narratives in terms of three overlapping themes: a vital personal connection to research; dilemmas of time and space to research and write; and stories of (more or less) successful navigation. She concludes that hopes for self, career, and service animate these teacher educators' stories and provide important reminders of the possibilities to participate in and develop professional life in ways that reaffirm "intellectual and spiritual growth, meaningful research and good relations" for all.

Chapter 12, by Judy Sillito, moves out of the university setting to focus on teaching English as a second language (ESL) in a center to aid immigrants and refugees in Canada. First, Sillito offers a personal narrative about the circumstances that have led her to research hope. She then discusses her research findings in terms of what she calls the four "natures of hope": the *existential*, the *relational*, the *realistic*, and the *transformative* natures of hope. These *natures of hope* are presented in light of how they relate to the ESL teaching/learning experience. Sillito also focuses on methodological and ethical issues to call into question how we might be better prepared to understand the dynamics of *transformative learning* in adult education. She finishes the chapter with considerations on her own personal transformative learning experience in the course of doing this "hope work" and how it continues to impact her personal and professional life.

In chapter 13, Yi Li, one of the two non-Canadian born authors who research hope included in this collection, turns the discussion back to the academic setting. This chapter is also a self-narrated story that seeks to record and understand the author's own learning experiences of becoming a teacher educator in a Canadian university. Herself an immigrant from China, Yi Li discusses, in particular, the challenges she has faced, the successes she has experienced, and her "growing sense of being and becoming a confident and effective teacher educator" to preservice English native speakers. In the *retelling* of her stories, hope emerges as a dominant theme. Yi Li also suggests ways to foster and support hope in new academic contexts.

Chapter 14, by Andrea M.A. Mattos, closes the collection and the stories of hope. Mattos showcases stories of hope from preservice EFL teachers enrolled in a course on language teacher education in Brazil. The author seeks to show "how reflecting on hope may help to provide language teachers and teachers-to-be with a sense of new possibilities and expectations that enhance their empowerment" in the profession they have chosen in face of the many threats the Brazilian educational context has recently posed. Mattos draws on the work of Denise Larsen (this volume) and collects a number of written narratives that reveal strong hope stories and different meanings for participants' hopes. She also discusses the presence of hopelessness in the preservice teachers' stories, but surprisingly finds out how these participants transform the threats to their hopes into more hope for becoming teachers. In the title of the chapter, the author plays with the similar sounds of the words *tale* and *tail*, as respectively referring to the participants' stories—or tales—and to their faint hopes for success in the teaching profession.

A criticism of narrative research has been that sometimes it comes across as content free. Some of the chapters in this collection, however, are strong examples of their content area research on teacher knowledge, offering insights on such diverse areas as science education, physical education, literacy, and foreign/second language teaching. The chapters in *Narratives on teaching and teacher education: An international perspective* describe the trials and triumphs of everyday teachers and teacher educators in their struggle to understand and live through the complexities and incongruities of educational practice. The wide variety of stories represented in this volume—from individual narratives of preservice teachers in several different contexts, to narratives of teacher educators and counselor educators, to team-teaching and faculty development—provide a varied account of the intricate paths of the teaching profession and educational settings. The stories also reveal the diverse possibilities for discovery and transformation of the individuals involved, and how hope and hopelessness may influence the choices and decisions that these individuals make. The narratives in this collection invite us to rethink and reflect on the richness and importance of our experiences and stories of teaching and learning to teach, and how these stories may impact the various contexts in which we work. Let us listen to these stories, then, and connect ourselves to the pulsing heart of the narratives on teaching and teacher education. Sit back and enjoy.

References

Bruner, J. 1986. *Actual minds, possible worlds.* Cambridge, MA: Harvard University Press.
———. 1990. *Acts of meaning.* Cambridge, MA: Harvard University Press.
———. 2002. *Making stories: Law, literature, life.* Cambridge, MA: Harvard University Press.
Clandinin, D.J., and F.M. Connelly. 2000. *Narrative inquiry: Experience and story in qualitative research.* San Francisco: Jossey-Bass Publishers.
Dewey, J. 1938. *Experience and education.* New York: Simon and Shuster.
Luke, A. 2004. Two takes on the critical. In B. Norton and K. Toohey (Eds.). *Critical pedagogies and language learning* (pp. 21–9). New York: Cambridge University Press.

CHAPTER 1

Narrative Frameworks for Living, Learning, Researching, and Teaching

Anne Laura Forsythe Moore

Prologue: Research, Like Water, Finds Its Own Level

The students who supported my research for this chapter were teacher candidates for whom I was course director in the Preservice Concurrent and Consecutive Programs of the Faculty of Education at York University, Toronto, Ontario, Canada, 2007–2008.[1] "Puzzle Framed" is an inquiry-based exercise integral for an Action Research component to curriculum planning, design, and implementation prepared for preservice candidates in parts I, II, and III. The exercise encourages the practice of viewing our world with plural perspectives in a more multidimensional way.

My position as a course director and instructor is that Action Research begins with a form of self-study, "an opportunity for self-critical inquiry undertaken by the participants themselves" (Holly, Arhar, and Kasten 2005, p. 31). I promote an awareness of wonder, a self-questioning curiosity, keen observation skills, and a passion to fit together the pieces of a particular personal research puzzle in an Action Research approach to curriculum. This is most often a niggling question that the candidate has at the onset of the introduction to part I: Action Research.

This process gradually evolves into the relationship of the curriculum to its "Commonplaces" (Connelly and Clandinin 1988, pp. 83–86; Schwab 1978, 1983): teacher/learner/subject matter, and social milieu. How the teacher designs and plans curriculum for the learners in the context of the social milieu and the subject matter becomes consciously founded on personal and professional experiences. Critical thinking and critical literacy skills are augmented through the initial autobiographical narrative inquiry. The enquiry of the candidates' Puzzle Framed leads to new knowledge of self.

Buoyancy with Data

I discovered as a coparticipant/teacher that to move like water does to find its own level encourages a fluidity of thought and buoyancy with research data that is fundamental to a deeper understanding of coparticipant stories. As I have said elsewhere, "so, like water, I move, I wind and I seep through the levels of stories touching surfaces and making connections to my knowledge buildings" (Forsythe Moore 2006, p. 133).

Active engagement in reading, writing, listening, and sharing stories of experience in a literacy community supports a personal and professional awareness that sustains a curriculum design that is democratic, equitable, and socially just. This phenomenon is vital in pedagogical approaches and policies that maintain in their overall structure a respect for the cultural influences that impact individual learning styles.

Candidates Choose a Teacher Metaphor

> Now re: your Teacher Metaphor: a flute. I feel it is my job to impassion with passion and no matter whether a trickle, a brook or a waterfall, the passion will re-present what it means to...savor, question, embrace, explore and to learn more...It is the great mystery of teaching. To light up for another how "it" works. Through love, really, and acceptance, understanding. Your music re-presents this and for this I would play my music as much as possible. My task as teacher is not to be concerned if the music is heard instantly, but rather that it lingers in the back of a mind until the right moment for the mind to go "aha." Sometimes we wait forever, sometimes [it happens] right away, sometimes next year—part of the role of teacher.[2]

As a result of an e-mail conversation with the teacher candidate to whom I sent the above response, the context for this chapter shifted. The curricular moment that is inclusive of the teacher's three-dimensional exercises of inquiry into her personal puzzles, those that have informed her and influenced her as a person and a teacher, requires "a reach beyond" self-inquiry during the enactment of curricular moments. Curricular expectations for each grade are judiciously laid out by the Ontario Ministry of Education, allowing our Ontario, Canada, teachers to speak a provincial language of expectations. The foundation is there. Everyone has the same framework to follow. The innovation with curriculum planning, design, and implementation of the curriculum, of course, is on an individual basis.

Sample Curriculum Outlined for Action Research: Consecutive Program for Teacher Candidates

1. Puzzle Framed: Part I: Action Research: A research question into self; framed and reflected upon from at least two different perspectives on the human landscape.
2. Personal Prospectus: Part II: Action Research: The teacher candidate's personal meaning for curriculum based on the relationship of curriculum with the commonplaces: Teacher, learner, subject matter, social milieu. The prospectus includes the candidate's teacher metaphor: What best describes the candidate's actions as a teacher?
3. Final Paper: Part III: Action Research: Understanding the commonplaces of curriculum through experience and education. The paper includes the candidate's present personal meaning for curriculum (personal prospectus), with reference to their teacher metaphor, observations, and "actual life experiences" (Dewey 1938) as they apply an attached lesson plan template to demonstrate how all these factors have sustained their curriculum planning, design, and implementation.

Continual Shifts: The Three-dimensional Narrative Inquiry Space

In narrative inquiry we name "understanding self" as the base, or the foundation of curriculum. The phenomena to examine in our preservice-narrative-focused curriculum that I have developed are the personal and professional stories of experience of the past and present social history of the candidates with a Next Steps component inclusive of a futuristic blend based on an "experiential continuum" (Dewey 1938, p. 33).[3]

My teacher candidates are encouraged to find curriculum anywhere and everywhere. We discuss how curriculum is located in their personal and practical experiences outside the classroom and in their professional experiences to date. The official curricular guidelines are used as reference points for skill development, meaning-making, and knowledge building. A strong focus on critical literacy, critical thinking, and critical autobiographical narrative inquiry is maintained.

The "three-dimensional narrative inquiry space" is "the direction this framework allows our inquiries to travel—inward, outward, backward, forward, and situated within place" (Connelly and Clandinin 2000, p. 49). The phenomena (*the what* of the inquiry) under study in a narrative inquiry are the stories (Connelly and Clandinin 2000, p. 4), and the method (*the how* of study) (Connelly and Clandinin 2000, p. 127).

> There is irony in this. The texts that speak most directly to the way in which we have begun to answer our question, that is, in terms of life, are the texts where it is most difficult to discern the answers we need. The answers are in the stories, indeed, the answers are the stories. (Connelly and Clandinin 1995, p. 79)

The Teacher/Learner Relationship: 3-D Inquiry Space

The teacher's story is important; therefore, the teacher candidate's story is important. In any teacher/learner relationship social histories overlap. It is the nature of human relationships when two components depend on each other for effective outcomes. The teacher/learner relationship could be compared to a hologram. I experienced a holographic effect when my teacher self was uniquely superimposed with my personhood—who I am as a woman. One purposeful outcome for narrative inquiry writing opportunities to be presented by the instructor is for self-expression based on social history. Personal interest is evident in the archetypal question, "Who am I?"

Within "the autobiographical narrative three-dimensional inquiry space" (Connelly and Clandinin 2000, p. 50),[4] I concretize what brought me to this work of exploration and discovery, "recovery and reconstruction" (Connelly and Clandinin 1988, p. 81).

Subtext for My Practice: A Literature of Self

An application through multiliteracies of the three-dimensional approach to storytelling sanctions a continuation of the enquiry into my own practice as a teacher. Through a rigorous reflection and analysis of my teaching practice, I began to isolate the strategic pedagogical underpinnings of my practice that have supported me as an Intermediate classroom teacher and then as an instructor in preservice programs. Through these stages of inquiry I consciously listened to my own voice. While hearing my teacher's voice, my writer's voice became stronger and more confident. My teaching practice had informed my own writing process as I reread for deeper meanings a "literature of self" (Forsythe 2004a), a term I created for a presentation on narrative inquiry that described my self-inquiry process during my doctoral journey. I allowed myself to free-fall with my personal experiences as a woman. I began to write poetry as a form for self-study and semi-autobiographical short stories to share with my students as part of our curriculum on the construction of the short story. I wrote articles for teacher journals based on my classroom experiences that were published and shared with my peers and colleagues in the field.

It is with intention that I create through curriculum design a similar opportunity for my teacher candidates to study a literature of self. The application of multiliteracies, including the teaching strategy of the Puzzle Framed, has made it possible. This self-study begins the inquiry into "who am I?"

The teacher candidate's e-mail opened my eyes to another phenomenon of knowing ourselves. This knowing continues throughout every teaching moment as we build on our teacher knowledge. I refer to the "knowledge structures" (Forsythe Moore 2006) we locate in our personal and professional stories as "readable texts" (Forsythe Moore 2006) for knowledge building and meaning-making. As researchers we must also be in constant movement like water and continually get our feet wet as the fluidity of our research curriculum transforms our directions. The inquiry needs an open door policy for new directions of movement along the experiential continuum of research, recognition, and discovery.

Now, back to the beginning.

A Renewed Narrative Perspective: A Knowledge Structure: "The Box" and Its Contents

"The Box," a cardboard container, was bequeathed to me by Annie, my maternal grandmother. It held various archival family documents including scrapbooks and letters dated from 1840–1980, the year of Annie's death. Its contents became my phenomena for research. It tracked the lives of early colonists, a blacksmith and his family who arrived in 1826 by ship to Pictou County, Nova Scotia, a Maritime Province of Canada on the Atlantic seaboard. My ancestors were colonists, of course. Confederation of Upper and Lower Canada (Ontario and Quebec) along with New Brunswick, Nova Scotia, and Prince Edward Island occurred July 1, 1867, several years after their emigration from Staffordshire, West Bromwich, England.

Long ago discarded by me, the imagined framework of The Box and its historical contents symbolized the frame and "narrative unity" (Connelly and Clandinin 1988, p. 59) of a varied social-historical past. Before I knew of its existence, the contents of The Box had been a reference point for my curriculum as a young girl growing up in postwar New Brunswick, Canada, that was passed to me through the oral tradition by Annie. She never mentioned the historical letters, yet always told the stories while I sat with her in the kitchen as she baked and scrubbed.

Location of Knowledge Structures:
One Methodological Approach

The Box framed a "social mythology" (Frye 1963, p. 67) formed from the historical information that had been carried forward by the ordinary language found on slips of papers, in letters, through mementoes, photographs, and entire scrapbooks. My curiosity led to questions that directed me down many roads to observe, inquire, explore, and discover. I began to understand how all the stories I gathered from family, past and present, regardless of gender, had become "knowledge structures."

The Box became a symbolic phenomenon to research. Teachings emerged from "the recovery, the reconstruction" (Connelly and Clandinin 1988, p. 81), the rewriting and reliving of the family stories. The frameworks of home and family and the actions of the people representative of their life experiences became significant evidence of the implicit and explicit knowledge structures that lie within family stories. The teachings of The Box encouraged me to recover, reflect, and to frequently reconstruct both my personal and teacher knowledge.

Its contents told and taught the story. But there was more. Within its contents my Puzzle Framed was located. The directions my research would take followed the road to "The Hingley Homestead."

Understanding our narratives develops a deeper empathy for the lives of the learners before us. I used many of the artifacts from The Box to model for the candidates how curriculum generally materializes from the teacher's own social history, personal interests, experiences, and passions, including the teacher's Puzzle Framed.

Personal Curriculum: Artifacts as Social History

Puzzle Framed

"Puzzles Framed" (Forsythe 2003, pp. 53–54) is actually a cameo collection of previously untold stories, those that had puzzled me for some time. The puzzles were part of my original proposal for my doctoral thesis. They had been shaped from my practice as both a teacher and a writer. They are intricate narratives unified in structure as part of the mystery that unfolds as the puzzles are pieced together through a variety of lens. The puzzles, or intricate narratives, address significant parts of my narrative inquiry. Why had these ideas, these personal family stories surfaced? What was their significance in my own narrative unity, in my life? I began to listen to the stories and gradually composed text from my reflections.

Teacher Knowledge

Throughout my inquiry a subtext of my teacher's voice as well as my practice as a teacher emerged. I explored notions that through my personal process of writing I could understand with more clarity the theory of how I might support my students to identify their own puzzles. I determined that my writing practice became a form of personal knowledge that could be applied to my professional practice as a teacher. Intriguingly, my practice as a teacher initially pointed me in the direction of the enactment of a narrative inquiry method in my writing. The powerful hologram-like bond between student and teacher relationships had shaped reciprocal learning experiences. A narrative-inquiry-designed writing practice in a preservice program provides a practical model and a foundation for the candidates' teaching of narrative inquiry writing.

A Puzzle Framed Exercise

The following Puzzle Framed is an extract from my personal research that had been reflected upon and analyzed. The impact of the exercise on my self-inquiry and critical narrative inquiry (inclusive of processes of critical thinking and critical literacy) induced me to incorporate the following exercise in presentations at conferences during my doctoral journey, as well as in my current curriculum for preservice teacher candidates.

In my curricular activities leading to the curricular expectation, I provide the learners with the following model extracted from my doctoral thesis (Forsythe Moore 2006, pp. 79–81).

Puzzle Frame #3

It is? A photograph of the original homestead, taken ca. 1900, complete with a hen running around the yard and snapdragons in bloom hugging the tired wood surface. It is stored in an old shoebox by an elderly couple who had been present (as children) at the wedding of my maternal grandparents in

Figure 1.1 The Hingley Homestead, Caribou, Pictou County, Nova Scotia, Canada, ca. 1930
Source: Author's Private Photo Collection

1916. During a Sunday drive that July 1992 along the Sunrise Trail in Nova Scotia, Canada, my mother and father located the old road into the property. We are welcomed with open arms into a home, which long ago, had replaced the original one (Forsythe 2003, p. 53).[5]

Hingley Homestead: Reflection #1, Autumn 2004

The photograph of my maternal grandmother's homestead, where she had lived until her marriage on July 21, 1916 had been stored in an old shoebox. The shoebox is located in a closet in their bedroom the day of our visit in 1992 by the property's present owners. Over tea its contents are shared with us. What a surprise to hear that the couple as young children had been at my grandparents' wedding in 1916. Ten years later (2002) as I stare at Annie's petit point scenes on my wall of a small white house nestled among colorful trees in fall and snow covered trees in winter, I realize that these houses look very similar to the house in that old picture the Grants, the present owners, had in the shoebox. "I was learning through my grandmother's eyes to view a landscape from different vantage points.

I was assimilating how to teach what I had learned" (Forsythe Moore 2006, p. 80).

Recover, Reconstruct, Retell, and Relive

While the sample writing is quite detailed for my personal Puzzle Framed, I offer the suggestion for the candidates to begin with three file cards. The first file card asks, and briefly answers, the question: *It is?* Many candidates placed an image of their puzzle on the first file card and on the back of the card wrote their explanation.

The second file card holds the candidate's initial reflection (*Reflection #1*) based on the facts of the story as the candidate recalls and recounts them (they had the option of using both sides of the card). The third file card contains a more detailed reflection that looks at the recounted story (*It is?*) and *Reflection #1* from another point of view, someone involved in the story yet not actually named in either *It is?* or in *Reflection #1*. For example, Annie is not actually named in *It is?* Yet she becomes quite significant in *Reflection #1*, which has moved the puzzle to an entire new level of thought. I learned something so intimate about my grandmother in "the recovery and the reconstruction" (Connelly and Clandinin 1988, p. 81) of my personal puzzle that without a narrative inquiry thinking and critical literacy approach I may never have recovered. I additionally encouraged the candidates to contact, visit, or interview a family member for more details regarding their puzzle to include in *Reflection #2*.

Varying Viewpoints from Past Time to Present Time Narrative Inquiry Research Methodology

The narrative inquiry into the mysteries of the quaint wooden structure of The Hingley Homestead was influenced by the suggestion that, "Narrative inquiry characteristically begins with the researcher's autobiographically oriented narrative associated with the research puzzle (called by some the research problem or research question)" (Clandinin and Connelly 2000, p. 41).

At this juncture of my experience with the methodology of narrative inquiry, I could reconstruct the research puzzle using a historical lens:

> How have my family's pioneer and early Canadian women's stories of immigration and migration, and of early settlement patterns in both rural areas and in towns, informed my identity as a Canadian woman and teacher on the landscapes of my birthplace, the Maritime Provinces

of Canada and of the province of Ontario, Canada, to where my parents with me had migrated?

I continued to puzzle:

> How has my education as a woman and a teacher been impacted by the family narrative unity that developed from the multiple accounts represented by historical documents, stories of relationships and narratives that I have termed knowledge structures?

As an outcome of my historical research I proposed that my life and education as a woman and a teacher were influenced by the family myths that were formed from the narratives, or "knowledge structures" represented by the historical documents found in The Box.

My candidates are encouraged to locate their personal boxes of information in whatever form they could be discovered. Multiliteracies as exploration formats are fundamental to my teaching strategies. This approach is intended to support the candidates in the curriculum planning for their classes in the future. When we experience a multiliteracy approach, we can teach it too.

A global research question emerged from my original research puzzle that further shaped my inquiry as a teacher. The question guided me and focused my intentions. I pieced sections of the puzzle with the application of an autobiographical narrative inquiry that included my coparticipants representative of a diverse demographic:

> How are narratives of unity created from narratives of pluralism within families and within communities that inform our personal and professional lives?

To understand the ways of life and the ways of thinking in one family is a good starting point. One generation of a family, carries unfathomable complexities. "Diversity is native to every household and countryside. Home is the heartland of strangeness; perhaps, then, we can learn to think of home as the best place to learn to live with strangers" (Bateson 2000, p. 13).

I was pulled into the inquiry of my pioneer past by the stories told through the oral tradition and through the historical materials found in The Box. I have an impression that the stories found in The Box in their varied forms carry a unity of thoughts and ideas although they represent two centuries of human lives. The quest to seek an understanding of what these unities might be initiated my intrigue with *the what* and *the how* of narrative inquiry. This research methodology considers the investigation of a family's stories through multiliteracy formats as a purposeful technique for gathering and displaying data as well as providing social historical evidence of what it means to be essentially human.

Is it probable that these stories as narrative unities, or "knowledge structures," have carried within them personal mythologies that were passed down through the generations? It is vital to explore through stories of experience these personal mythologies. They are tools for self discovery. We teach who we are. To determine through stories the influences that have informed our education as human beings helps to ground us in our gender while it sustains a curriculum design that supports our students' social historical exploration and research. It also opens stories for further inquiry and reflection in the present time that have been stuck in one frame from generation to generation. The best part of this process is that through this work, high-level skills are developed.

Dear Teacher,
You have had such a large impact on me and given me the opportunity to inquire about family genealogy, my childhood experiences, and my beliefs and understandings of what curriculum means to me as a teacher learner. You've taught me to have a point of view on my point of view and you have made a difference in my life and I thank you for that.[6]

A Fluid Inquiry

A Critical Analysis of Personal Puzzles

This reliving of a past is a commitment to live with mindfulness and discernment. Most candidates are surprised about what they have been wondering and why they have not spent more time critically analyzing their personal puzzles. The Puzzle Framed exercise integrated in a preservice program provides an opportunity for the candidates' self-exploration and "hands-on" learning for how the process works and how significant the discoveries can be.

Connelly and Clandinin (2000, p. 184) characterize "narrative inquiry as a kind of fluid inquiry...that necessitates ongoing reflection, what [they] have called wakefulness." I accepted this challenge. I now have passed the challenge forward to my candidates. Through reliving the experiences by the artifacts set before me such as the photograph of The Hingley Homestead, I try to tell the story well. I constantly share my stories, my puzzles, my reflections, and discoveries with my candidates. And, they are encouraged to share their stories. I find it intriguing to observe how many frames include puzzles about houses of the past—frameworks of social mythologies. There are always a variety of puzzles depending on the size of the class, however, and many candidates take the time to send me personal reflections:

> I am sure that most of us will continue the search for solving the "puzzle framed," extend the inquiry into our own genealogy, and engage our own students into similar activities for the future. Your insight has inspired us to go way beyond what was required.[7]

New Knowledge: Levels Professional Practice

Research Moments: Stories as Data

Enter candidate L and candidate G. When I asked my teacher candidates if anyone would submit their Puzzle Framed, for this chapter, candidate L immediately offered hers. Something intangible in candidate G's, something sitting below the surface, something present, yet something missing made me ask her if I could use her Puzzle Framed. Also, I found through the reading of candidate G's inquiry recorded in her Puzzle Framed that I experienced a hologram effect with the overlap of my narrative inquiry into the significance of Annie's attention to her petit points of the four seasons. It is the holographic experience that sustains the teacher's commitment to the learner's story; the researcher's data that sustains her commitment to the research. There is something so fundamentally human in the holographic experience.

With a critical narrative inquiry approach to her Puzzle Framed, candidate G has been very reflective toward the experience in general. She learned that the reflection on the inquiry is in the control of the participant. "You can allow your reflection to go only as far as you want it to go" (Expressed in a group conversation during our "Communications and the Education Process" class, Thursday May 15, 2008). We have presently reached that space.

Candidate G varied my suggested format for the Puzzle Framed inquiry. Candidate L followed my format model.

Candidate G's Puzzle Framed

1. Why do I love red so much?
2. I wonder if it's because the carpet in our home growing up was red. Maybe it's because we had a red couch and matching armchair that my

mother sat in when she knitted... Maybe it's the smooth leather armchair we had when I was about 11 years old. Maybe it's the red skating outfit my mother sewed for me just before she died. Maybe it's because everybody tells me how good I look in red.

Candidate G originally submitted her Puzzle Framed with her question neatly keyed on the first file card that she labeled *1*. The above entry on the second file card was labeled *2*. The third card reads:

> 3. "Red is Best," written by Kathy Stinson and illustrated by Robin Baird, is an old book that has come back into fashion. I LOVE the little girl in this book. Two things she says stand out for me. They make me feel really good! Here they are: "... my red barrettes make my hair laugh" and "... red paint puts singing in my head."

Coresearchers' Stories Overlap

My relationship and experiences to Annie's petit points had brought me closer to the realization that Annie had been stitching the decades of her childhood and early adulthood memories together as her nimble fingers worked with the needles and the wool. There are multiple ways to tell stories.

While candidate G and I have barely entered the coresearcher/coparticipant relationship, as I read her Puzzle Framed I queried the absence of *Reflection #2* and offered some suggestions. The following is my response to candidate G's *Reflection #1*, written on February 25, 2008.

> What are the thoughts in a human's mind when they choose the color red for...
>
> Red is such a statement in ♥ month, isn't it—Red is for Heart, for Valentines, for love, for passion, for strength, for hope... red barrettes make my hair laugh... red makes... "wonderful lines."

I asked candidate G if she would continue to reflect on her Puzzle and I gave her an opportunity to resubmit, which she did with her *Reflection #2*, a friend's point of view on why candidate G loved "red."

Teacher Knowledge for Curriculum Planning

It is of interest to note that our stories are founded in our literacy roots and cultural identity. As teachers we bring our heritage to our classes as a teaching tool. My heritage, my cultural roots, is who I am. It is a *literature of self*, familiar curriculum with which to plan and to support our students' self-appreciation of their stories of experience.

It is my intention to demonstrate that each and every teacher candidate has a *literature of self* to bring to their curriculum planning. A pedagogical practice that applies a narrative inquiry methodology to the teacher's own social history encourages the classroom protégé to follow suit. This models a democratic approach for curriculum planning to support candidates to envision the locations and spaces of their experiences simultaneously as they experience the sample curriculum built around my narrative inquiry into the puzzle of The Hingley Homestead.

Hingley Homestead: Reflection #2, April 2005

"My mother told me that once Annie, her mother, bought a ribbon for one of the little neighborhood girls. I have often wondered if that little girl had been Mrs. Grant" (Forsythe Moore 2006, pp. 80–81).

Further Reflection: #3, Winter 2008

My students and I experienced what it means to view a story from a variety of points of view, a methodology essential in social science research. Sometimes a ribbon makes a difference in a child's life. How might this knowledge shift our perspectives as a teacher in any classroom in any part of the world?

Candidate L's Puzzle Framed

It is? It is a photograph of my Great Uncle.

This picture was taken in 1948 before he served in the Korean War. He was only 18 when this picture was taken. It is kept in my grandmother's purse wrapped up in a clear plastic bag and has been there for as long as I can remember. He was the oldest of seven; however he did not grow up with my grandmother. He had lived with his grandparents, as my great grandma was too young to take care of him…

Candidate L had offered her Puzzle Framed very quickly when I asked the class in general. She offered it so perhaps she is open to further exploration that will be ongoing this year 2008–2009. Candidate L's photograph created a hologram effect for me as I superimposed my father's picture taken "Somewhere in England" after December 8, 1939, when the 8th

Field Battery, 2nd Field Regiment, First Canadian Division of the Royal Canadian Artillery, landed on British soil. Candidate L and I lost our fathers recently.

Personal Curriculum: A Tool for Recovery of Meaning

Knowledge Structures: Readable Textbooks

It is with humble inquisitiveness that I read the candidates many puzzles that represented the significant diverse perspectives of a culturally pluralistic demography.

The teacher's personal curriculum as a teaching tool illustrates for candidates that stories of experience are significant. It is a win-win situation when learners engage with their own curriculum. Of course, eventually the next stage of personal development takes the learner outside of their own stories to the stories of others. This process enhances personal growth and development of the learner toward a more pluralistic world view. The sharing of stories in a community that promotes a level of comfort for sharing leads directly back to a personal transformation and a multidimensional perspective for the learner with respect to others.

The homestead and other spaces of my childhood, youth, and adulthood became "readable textbooks" (Forsythe Moore 2006) that were the foundation of my personal curriculum. I questioned how it could be considered that culturally pluralistic environments are only those environments, which illustrate a diversity of cultures, without the consideration of diversity of regions. I explored notions that one family can represent a global village and one classroom can sound the heartbeat of a nation. If the structures of home and the family stories within its frame have become readable textbooks of diverse family stories foundational for my personal curriculum, then the homes of others can be readable textbooks as well.

The Power of Reflexivity

The storied experiences of our students integrated as part of curricular development for preservice candidates provide the groundwork for multiple discoveries of the meaningful connections between living, learning, researching, and teaching. Sharing and listening to the stories of others at home, in the classroom, or in the community at large, have affirmed for me the power of reflexivity in the teacher/learner relationship and how important this understanding is in the field of teacher knowledge, education for teachers, and curriculum development.

An invigorated understanding of a diverse cultural perspective on humanness and human stories reflects the philosophy that a developed personal and professional awareness supports a democratic curriculum that is equitable and socially just.

Canadian Perspectives through the Four Seasons of Life

The synthesis of my research into this text illuminated the meaningful connections between living, learning, researching, and teaching. My present self became more real because of what I was able to see through the scrim of holographic shifting between the past and the present of my own puzzles, particularly of The Hingley Homestead as depicted through the seasons by Annie's lovingly crafted petit points and those of my students.

Through the writing of this chapter, I completed a full circle to connect my ending with my beginning. It takes one person at least to remember the story. It takes one person at least to tell the story their way. And, it takes one person at least to teach the story by living it. Most importantly, however, it takes one person to believe the story enough to question it. To be an agent of purposeful change in our own ways of thinking for the public good is a positive human construction.

Can we become our curriculum story for others, yet continue to be a seeker ourselves while sustaining an open-ended inquiry-based curriculum? This, of course, is the nature of narrative thinking.

Notes

1. The candidates in both Consecutive and Concurrent Programs have completed four years of an undergraduate degree and will have a Bachelor of Education on the completion of their preservice courses and practicum blocks.
2. This e-mail was sent to a teacher candidate (April 17, 2008, 8:37pm) upon receipt of her e-mail: "I have come to realize that just because I am passionate about something, it doesn't mean that all my students will be too!"
3. Experiential continuum: "Every experience both takes up something from those which have gone before and modifies in some way the quality of those which come after" (Dewey 1938, p. 35).
4. I use a shortened form, "3-D," in assignments and rubrics and refer to the 360 degree perspective as the analysis of stories from all angles and from varying points of view.
5. After this had been published and during subsequent research for my thesis, it was determined that the photograph had been taken after 1930.
6. Letter from teacher candidate, July 11, 2007.
7. E-mail reflection from a preservice candidate, Wednesday, July 18, 2007.

References

Bateson, M.C. 2000. *Full circles, overlapping lives.* New York: Ballantine Books.

Clandinin, J., and M. Connelly. 2000. *Narrative inquiry: Experience and story in qualitative research.* San Francisco: Jossey-Bass Publishers.

Connelly, M., and J. Clandinin. 1988. *Teachers as curriculum planners: Narratives of experience.* New York: Teachers College Press.

———. 1995. Narrative and education. *Teachers and Teaching: Theory and Practice* 1(1), 73–85.

Dewey, J. 1938. *Experience and education.* New York: Simon and Shuster.

Forsythe, A. 2003. Puzzles framed. *Among Teachers: Experience and Inquiry* 33, 53–4.

———. 2004a. *Narratives of storied student-teacher relationships: Seeds for learning.* Paper copresented at the Ontario Education Research Council (OERC), Ontario Institute for Studies in Education, University of Toronto, Canada.

Forsythe Moore, A. 2006. *A teacher's stories of living and learning: A narrative inquiry into the relationship between family, home and education.* Ph.D. diss., Ontario Institute for Studies in Education, University of Toronto.

Frye, N. 1963. *The educated imagination.* Toronto, Ontario, Canada: House of Anansi Press.

Holly, M.L., J.M. Arhar, and W.C. Kasten. 2005. *Action research for teachers: Traveling the yellow brick road.* 2nd ed. New Jersey: Pearson Prentice Hall.

Schwab, J. 1978. The practical: Translation into curriculum. In I. Westbury and N. Wilkof (Eds.). *Science, curriculum and liberal education: Selected essays* (pp. 365–83). Chicago: University of Chicago Press.

———. 1983. The practical 4: Something for curriculum professors to do. *Curriculum Inquiry* 13(3), 239–65.

PART I

Stories of Discovery and Transformation

CHAPTER 2

Understanding Classroom Experiences: Listening to Stories in order to Tell Stories

Andrea M.A. Mattos

The classroom is a microcosm in the sense that what happens within the classroom reflects, affects and is affected by the complex of influences and interests within the host educational environment.
—Holliday 1994, p. 16

According to the developmental psychologist Jerome Bruner (2002), narratives and stories are a way of thinking, a way of organizing human experience. In the author's view, plots that have a beginning, middle, and end provide us with frameworks that contextualize the information we process (Barry 2006). Bruner (1990, 2002) also states that it is through narratives and stories that we construct an image of ourselves—it is through telling and listening to stories, including our own, that we are continually formed and transformed, and we become who we are.

Fivush (2006a), a psychologist who researches family narratives, says that "narratives are the way in which we make sense and create meaning from our everyday experiences, and this process occurs within social interactions." She states that "as we talk about our experiences with others, we reinterpret, re-evaluate and reconstruct our experiences for ourselves." The author defines narratives as "culturally constructed ways of understanding what a life and what a self is" (Fivush 2006b). She later adds that

> Narratives move beyond the simple description of experienced events to provide explanatory frameworks and emotional evaluation of what these events mean to the individual. More specifically, narratives allow us to create a shared reality. Through telling the stories of our lives, we are telling who we are and we are sharing our view of the world. We do not simply tell what happened; we explain how and why these events occurred, how we thought and felt about them and what they mean to us. (Fivush 2006b)

The area of language learning and teaching has lately been interested in using narratives as research methodology. One of the long-lasting challenges in the field of foreign language learning (FLL) has been understanding and explaining how language learning takes place. Narratives have recently been used as a way of getting access to crucial information that would, otherwise, be inaccessible to the researcher. Pavlenko (2001), for example, believes that learning memories have great potential to second language learning (SLL) research. She says that narratives are "a unique source of information about motivations, experiences, struggles, losses and gains" (Pavlenko 2001, p. 213). In the area of teacher education, Carter (1993) and Clandinin and Connelly (2000), focusing on teacher knowledge and development, are major references.

This chapter reports on the results of a study that investigated a teacher of English as a foreign language (EFL) and her perceptions of her own classroom. The main objective was to better understand the foreign language classroom and the participants' experiences from the teacher's point of view. In order to have access to the teacher's perceptions and interpretations, narratives, collected in the form of retrospective interviews, were analyzed. Before we get into the description and discussion of this study, let us know more about the "stories" told by language teaching research.

A Short (Hi)story of Research in Language Teaching

In a seminal article on language classroom research, Long (1980) discussed the methodological issues involved in this type of research and claimed that the language classroom is a "black box," in an allusion to the scant knowledge obtained about classroom teaching/learning processes and to the almost total absence of research in classroom contexts at that time. Although this is now changing, what is known about the foreign language (FL) teaching/learning process and about the classroom as a social context is still very limited and tends to adopt the point of view of the learners (Freeman 1996a, 1996b), leaving a gap in relation to the point of view of the teacher as a participant in this context. Freeman (1989) and Richards and Nunan (1990) mark the beginning of a new trend toward filling this gap. Freeman (1996a, p. 360) claims that "there has been the need to study, to understand, and in a sense to define, *teaching* independent of its outcomes; this includes coming to understand the role and person of the teacher."[1]

In the 1990s, research on teacher education and development evolved quickly and went through several changes in perspective. Freeman (1996b) discusses the relationship between research and teaching and highlights its hierarchical and unidirectional nature. He argues that "you have to *know the story* in order to *tell the story*"[2] (p. 90) and calls our attention to a central dilemma in the field of FL teacher knowledge: the eternal divorce between research and teaching, since those who deeply know the classroom (teachers and learners) rarely talk about this knowledge, whereas those who are normally very keen on talking about the classroom (researchers), "often miss the central stories that are there" (Freeman 1996b, p. 90). The author suggests that

> To bridge the gap and to fully understand teaching, we must take an approach which puts the person who does the work at the center.... What teachers know, and how that knowledge finds its way into their practice, must become a vital concern of those who want to understand and to influence education. (Freeman 1996b, p. 90)

Freeman (1996b) groups research on teacher education into three different trends, depending on the point of view assumed by the researcher and the importance of the contributions of the participant teacher: the behavioral view, the cognitive view, and the interpretivist view. The behavioral view, as the name suggests, focuses on the behavior of the teacher inside the classroom. In this view, the researcher is the observer of what teachers do in the classroom and teachers' actions are frequently related to what students learn as a result. However, the results of this type of research are "detached from both the world in which [teaching] takes place and the person who does it" (Freeman 1996b, p. 91). The problem lies in the fact that this trend views teaching as "doing things" and "tends to codify extremely complex processes... simplifying teaching by not attending to the role that teachers and learners, as thinking people, play within it" (Freeman 1996b, p. 93). The results of this type of research, thus, in terms of the "stories" that are told, lead to a compartmentalized view of teaching, which is behavioral, impersonal, and detached from the contexts in which it occurs.

The cognitive view focuses on the teachers' mental processes, their perceptions and intentions, their beliefs, knowledge and attitudes, as well as the affective dimensions (such as feelings of anxiety and fear) that are undoubtedly present in their daily practice, shaping their thoughts and actions. This view includes "research on teacher cognition" or "teacher thinking" (Woods 1996), which tries to relate what teachers think before lessons ("pre-active decisions") to what they think during lessons ("interactive decisions"). Freeman (1996b) says this type of research defines teaching as "thinking and doing" and cites Halkes and Olson (1984) to exemplify the new research interest:

> Looking from a teacher-thinking perspective at teaching and learning, one is not so much striving for the disclosure of *the* effective teacher, but the explanation and understanding of teaching processes as they are.... Instead of reducing the complexities of teaching-learning situations into a few manageable research variables, one tries to find out how teachers cope with these complexities. (Halkes and Olsen 1984 according to Freeman 1996b, p. 95)[3]

This type of research tries to understand not only what teachers do inside the classroom, in terms of behavior and action, but also what they think while performing such behaviors and realizing specific actions. Freeman argues that "to tell this side of the story we have to place teachers' perceptions—their reasoning, beliefs, and intentions—at the center of any research account" (Freeman 1996b, p. 95). The author says that the results of this research type tend to reveal not only visions of teaching but also the limitations that frame it, and concludes that "teaching is not simply an activity that bridges thought and action; it is usually intricately rooted in a particular context" (p. 97).

Aiming to understand how teachers interpret their own practice in relation to the specific context they work in, Freeman suggests an alternative trend, the interpretivist view. He defines teaching and teacher development as "the unstudied problem," especially in the field of FL teaching, "in which traditional practices, conventional wisdom, and disciplinary knowledge have dominated" (Freeman 1996a, p. 374). This new perspective tries to view teaching as "knowing what to do" and intends to cater not only to the researcher's conclusions but also to the voices of the participant teachers. The author states that "such research provides a mirror that reflects the teacher centrally in [its] account" (Freeman 1996b, p. 99). Freeman concludes that teachers' professional knowledge is characterized as stories—"stories of knowing what to do." He contends that research on teaching should involve "knowing how to tell the story" (p. 99) in order to better interpret teaching and its outcomes.

The Nonnative Teacher

One of the persisting concerns of researchers in the field of language teacher education is the issue of the nonnative language teacher. The dichotomy between native and nonnative speakers[4] of a language dates back to the origins of the studies in Modern Linguistics, when Chomsky (1965) introduced the notion of "native speaker competence" (Stern 1983). Initially, the need to imitate the native speaker's performance dominated the goals of FL teachers and learners as well. Nowadays, this is no longer as important (Koike and Liskin-Gasparro 2003; Kramsch 2003; Mattos 1997), although the issue hasn't been totally dismissed.

Almost thirty years after Chomsky, in a book called "The Non-native Teacher," Medgyes (1994) reopened the discussion around this controversial aspect of language teaching. The author discusses the main differences in attitude between native and nonnative teachers of English as a Foreign Language (EFL), and the advantages and disadvantages of being a nonnative teacher. He gives special emphasis to the main problems faced by nonnative EFL teachers, such as the language deficit and an inferiority complex toward native speakers of the language. As a consequence, Medgyes states that the teacher may enter a stress cycle that may lead to decreased performance and bad feelings of underachievement. The work overload may also contribute to raise teacher stress, causing tension and feelings of self-blame, which can lead the teacher to isolation or even cause health problems. Medgyes (1994, p. 41) claims that

> Studies of the language learner are abundant, while those on the language teacher are much less common. This also applies to research on personal stress in ELT [English Language Teaching]: whereas books and articles on anxiety in language learning are in abundance, there is hardly anything written about "the sickness to teach" foreign languages. This is a regrettable fact, considering that anxiety-ridden teachers are likely to raise the students' anxiety level too. Learning about how to alleviate our own stress is a precondition for being able to deal with learner stress.

Medgyes (1994) warns us about the dangers of teachers becoming stressed and anxiety-driven. He says that understanding the nature of the problems encountered by teachers and clearly identifying the sources of teachers' anxiety and stress is mandatory in order to better prepare teachers to face these problems and to learn to avoid anxiety and stress in their daily practice.

Understanding Classroom Experiences

The research design for the preparation of this chapter was based on Freeman's (1996a, b) interpretivist view of research on teacher development, and adopted an emic vision in relation to data collection and analysis, that is, the methodology tried to focus on the participant's interpretations about her classroom events from her own point of view.

Carolina, my coparticipant in this research project, was an English teacher at CENEX,[5] a language institute[6] in Brazil. She was a novice teacher, still in her last year of the undergraduate program in teaching at the Federal University of Minas Gerais (FUMG), and she had no previous teaching experience. The context for data collection was one of her pre-intermediate adult groups at CENEX.

The research data was collected during the second month of the course, through a series of retrospective interviews, when Carolina reflected on her classroom events with the help of a video recording of her classes. During the interviews, Carolina provided stories and interpretations of the events in her classroom. The transcription of the interviews gave rise to a full list of reflective categories that provided a general view of the participant's perceptions of her classroom.[7] A detailed qualitative analysis of these categories revealed the stories told by the participant about her teaching experiences. These stories and experiences were used as the main data for this study.

Carolina's experiences and stories included not only her beliefs about teaching and learning, but also other perceptions and interpretations about various subjects and events. However, a detailed qualitative analysis of all her perceptions and interpretations would be extensively long and beyond the scope of this chapter. Among all the reflective categories, then, dominant ones were identified and selected for this analysis.

One of the dominant categories was labeled "the context" and represented a picture of the elements that influence the FL teachers' daily life, as perceived by Carolina. As discussed in Medgyes (1994), Carolina's stories seem to corroborate the main problems faced by nonnative teachers in their pedagogical practice.

Carolina's stories and experiences reveal that her classroom practice is influenced by various factors. This influence may be positive or negative and may originate inside or outside the classroom. In order to better represent the existence of these several factors, her stories and experiences included in

Table 2.1 Subcategories for the Participant's Dominant Stories

	The Context	
• The relationship between teaching and the context	—Positive influences —Negative influences	
• The forces inside the classroom	—Internal pressures	a) pedagogic pressures b) interactional pressures
	—External pressures	a) personal pressures b) contextual pressures

Source: Author's Illustration

the category "the context" were further divided into subcategories. Table 2.1 represents these subcategories.

These categories represent an overview of the factors that influence Carolina's classroom. As Medgyes (1994) highlighted, the nonnative teacher's daily practice is full of influences and pressures that don't always originate inside the classroom. Carolina talked about these influences and pressures during the retrospective interviews. In her narratives, Carolina told stories of her interpretations of the events in her classroom. Besides, she also talked about the reasons she attributed to each of these events. Her reasons include, for instance, her personal difficulties as a learner of the language she was already teaching, the positive and negative influences of her students' previous learning experiences, her difficulty in timing classroom activities, her feelings of self-demand, her desire to make a good image for her students, her difficulties and anxieties toward her other private activities and toward her pedagogic duties, such as the evaluation system and the need to fulfill the course program on time. The next sections will discuss these perceptions through examples that illustrate each of the subcategories in Carolina's stories.

The Relationship between Teaching and the Context

Positive and Negative Influences

Carolina highlighted, for example, as a positive influence of the context on her classroom, the issue of the teacher-learners relationship. According to her point of view, the image that the students from CENEX have of their teachers is different from the image of students from other language institutes. She said that, at CENEX, students are more tolerant toward their teachers' language mistakes because they know that the objective of the institute is to offer practice opportunities to preservice teachers who are undergraduate students at the university. In Carolina's opinion, this makes the teacher-learners relationship easier. The following excerpt[8] from the second interview illustrates her perception of this issue. She was talking about a mistake that she had made during the class. As I asked her how students react to teacher mistake, she explains her views on this issue:

> *R*: How do you think the students react when they notice the teacher makes a mistake?
> *P*: Oh, I think that…they are used to it…because they know…the teacher is not perfect. They are used to it. For example, if they have a question, and sometimes you don't know the answer, you have this option, this openness to say "I don't know, I'll bring the answer next class. I'll try to find out for you."
> *R*: And do you think this is something that normally happens to students nowadays, I mean, do you think they have this view that the teacher is not perfect?
> *P*: No, I think this is a characteristic of the students from CENEX. 'Cause, at other schools, I'm not certain, but I think at other schools they don't have this vision, they expect everything from the teacher (…) At CENEX, no. At CENEX they are more open to this issue…because since their first stage it's like this. So, they know that the teacher may also make mistakes, the teacher is not perfect.

At a subsequent interview, Carolina talked about the low level of students' participation during lessons, an issue that was identified as a negative influence of the context on her classroom. She mentioned that her students participated very little in classroom oral activities because, in their previous stages at CENEX, their participation wasn't systematically required. Her opinion on this issue is exemplified below:

> *P*: Well, they don't participate very much, do they? Here I think the problem is their previous stages…because they are already in the 4th stage, but maybe in their previous stages…I don't know…may be because of the short time…the teacher, in the other stages, isn't very demanding, right, and doesn't require much speaking from the students, doesn't ask them to repeat…or doesn't give them many speaking activities…I think this may jeopardize the students' oral skills in future stages.

The Forces Inside the Classroom

Internal Pressures

During the interviews, several times, Carolina touched on the issue of pressures in the classroom, be they of internal or external origin. Among the internal pressures, her comments either referred to pedagogic pressures or interactional pressures. In terms of pedagogic pressures, Carolina's stories highlighted two main points: the issue of classroom timing and unpredicted problems. The excerpt below, when she was talking about a time she had been very anxious, exemplifies one of the times she talked about the problem of timing:

> *P*: (...) So, I forgot to ask the questions, I forgot to make this discussion.
> *R*: So it was because of anxiety. Anxiety about what?
> *P*: Because of the time. This group is very...time is never enough for everything I plan...(laughing) So...I didn't even think of the "Grammar Focus" [an item in the course book], I simply went on, and...I forgot that.

The next excerpt shows one of the times Carolina touched on the issue of unpredicted problems. During this interview, she was talking about the difficulties that students usually have in understanding grammar points that are different from their mother tongue, such as the concept of uncountable nouns. In her point of view, the teacher should predict these difficulties or problems and get prepared to clarify students' questions. However, when this doesn't happen, the teacher faces an unexpected situation, which was exactly what she had experienced:

> *P*: What is, I think, most important to me, as a teacher, is thinking about this issue, when there is, I mean, something that...we are not very much used to, right, about the language [English]. I think it is important to...predict the problems...And this is difficult, but you have to go to classes more or less expecting something. And sometimes there are things you never expected and this is very difficult...I think that's what is most difficult: you have to know how to deal with unexpected problems, right?
> *R*: And has this ever happened to you? I mean, have you ever predicted problems that really came up, or have there been problems that you never predicted?

P: There have been problems that I had already predicted. And problems that I never predicted. On this confusion about countables... because "traffic" is uncountable, but... [the student] compares with Portuguese and thinks it's countable. This I had already predicted and it really happened. But, thinking that an adjective may be countable or uncountable, this I could never... imagine!

In relation to the interactional pressures, Carolina showed that she was concerned about her students' image of herself. She mentioned that a teacher should show assertiveness to the students and her desire to fulfill this expectation worried her, as it may be noticed in the excerpt below:

R: Here [referring to the video], what happened? You translated some words, like "peas" and "frozen." And then you said "sorry" because of that [laughing].

P: Right [laughing], because I think my students like when I speak English all the time. I have already noticed that. And... sometimes they may think that translating may be... I mean... the teacher may be a little... lazy, or finds it [the word] difficult, and translates right away. So I said "no, [laughing] this is difficult." That's why I said "sorry." Because... that's all about expectations, right? [pause] I think there is no other way. We are always expecting something... mainly from teachers. If you are going to start a new job, you expect something. If you are going to start a lesson, you also have expectations. So, I think students have some idea, especially about the teacher... I think you always have that expectation in relation to the teacher, right? You want to see the teacher... at least you have to see that he[she] is an assertive person, that he[she] knows what he[she] is doing.

R: And that's what you want to show to your students?

P: Exactly... I think that if I show them assertiveness, show them that they may... have the environment, everything available, nice things to... to facilitate their learning, I think this... This is what worries me, because... before I can capture what their expectation is... I need to have this as a previous understanding... so that they may accept or not, right, and I may adapt.

External Pressures

In her stories, Carolina also talked about external pressures to her classroom. These were either personal or contextual pressures. The personal pressures were identified as those that related to Carolina's personal life and the contextual pressures were mainly those related to the school context.

When talking about personal pressures, many times Carolina mentioned the existence of problems related to her other activities, which would make her behave inattentively, or mindless, during lessons. One of the examples is represented in the following excerpt:

> *P*: This lesson wasn't good. Because I was extremely overloaded with other things. That day I was full of things to do. And I couldn't finish everything before the class. So I was worried about the things I had to finish for the class I was going to have at the university in the evening, after the class I was teaching. So I was very worried. I had to finish a paper to hand in that day, so I...was worried about that. And I saw how this disturbed my class! I was tired...I stumbled on the things I was saying, I was really...because I was anxious about my other activities.

In relation to the contextual pressures, that is, those related to the school context, Carolina talked about the problem of fulfilling the course content:

> *P*: ...I think I'm behind schedule in relation to the content of the stage, I mean, the other 4th stage groups.

Another issue she mentioned in terms of the contextual pressures in her classroom is related to the school evaluation system. For her, summative evaluation seems to be a necessary evil:

> *P*: Oh, evaluation...is very complicated. I don't like it. I don't fit in with evaluation. I agree that "tests don't prove anything." But...in the course you need something to grade the students, otherwise...what are you going to do about grading, these formalities, I mean, the bureaucracy, whatever? So, I think it is very boring. Having to evaluate students is terrible, I don't like. This also bothers me, when I have to think about something to use, because I know it will be used to evaluate.

These are examples of some of the comments Carolina made about the influences to her daily life as a teacher and about the forces in her classroom. Her stories represent the vast universe of influences and pressures to the classroom, which was discussed by Medgyes (1994).

The several elements that influence the FL classroom mentioned in Carolina's stories are represented in figure 2.1.

In figure 2.1, the star-shaped area represents the context in which the classroom is placed—the school, the community, the country, and ultimately

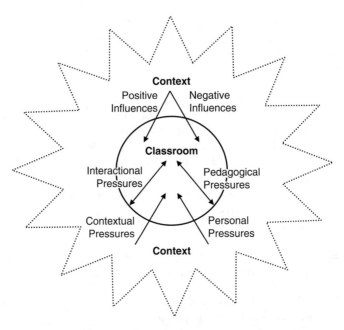

Figure 2.1 Elements that Influence the FL Classroom
Source: Author's Illustration

the world. Because it is fluid and unlimited by nature, the context is represented by a broken line. The several tips of the star aim to represent the contextual differences present in the various FL teaching situations. The internal circle symbolizes the FL classroom, which is represented by a solid black line because it is a closed culture, which, according to Holliday (1994), is not accessible to the outside observer or a "black box," to use Long's (1980) term. From the context, the positive and negative influences to the classroom emerge, as well as the external pressures, that is, the personal and contextual pressures. Inside the classroom itself, the internal pressures are raised, and take the form of pedagogic and interactional pressures. Carolina's stories ratify Medgyes' (1994) ideas when he says that nonnative teachers may face several problems in their practice.

Conclusion: Knowing the Story; Telling the Story

Through her stories and experiences, Carolina, the participant teacher, revealed that her classroom practice was influenced by a number of different elements, which may be either a positive or a negative influence or may take the form of pressures originated inside or outside the classroom. For example, as one positive influence on her classroom, she mentioned how the students' previous (good) learning experiences may favor the student-teacher relationship and, as a negative influence, how the (bad) learning experiences affect the teaching process. The classroom pressures were subdivided into pedagogic pressures (such as timing and unpredicted problems), interactional pressures (such as students' image of the teacher), personal pressures (such as self-demanding and the teacher's other private activities) and contextual pressures (such as the evaluation system and the need to fulfill the course program).

Clearly, teachers do find problems and other sources of anxiety in their daily life, and these elements influence the teachers' practice. Obviously enough, in order to have greater knowledge of the FL classroom and to understand it as a social context in all its complexity, researchers in the area of language teacher education will need to rely on teachers' interpretations and perceptions of their classroom events. Narratives and stories are an important tool in this direction and, therefore, placing the teachers' stories and interpretations on the central focus of our research is a necessary step for those who study teacher education. As Freeman (1996b, p. 90) stated, we "have to know the story, in other to tell the story."

Notes

1. Emphasis in the original.
2. Emphases added.
3. Emphasis in the original.
4. Lightbown and Spada (1993, p. 124) define the native speaker as "a person who has learned a language from an early age and who has full mastery of that language". In contrast, the non-native speaker is the one who has learned the target language after childhood and whose command of the target language presents limitations.
5. CENEX stands for *Centro de Extensão*, which is a university-based language center where foreign language courses are available to the community in general. The teachers who work at CENEX are mainly undergraduate language students from the Federal University of Minas Gerais, Brazil—where the study was conducted. It is at CENEX that these student teachers take their preservice internship as a requirement of their course.
6. In Brazil, English is a compulsory subject at regular schools for students at the age of eleven on. Besides, there are numerous language institutes where learners can study one or more languages at their choice. There is an outstanding number of such institutes that offer English to learners of all ages.
7. For reasons of time and space, these categories will not be discussed here. For a deeper discussion and a better understanding of the concept, please refer to Richards (1998) or Mattos (2000, 2002).

8. "*R*" stands for *researcher* and "*P*" stands for *participant*. Within brackets, [ ], are comments made by the researcher to facilitate understanding; a pause or hesitation is represented by…and (…) represents a longer pause. All the excerpts have been translated from the original.

References

Barry, H. 2006. *Jerome Bruner*. Retrieved January 17, 2006, from http://evolution.massey.ac.nz/assign2/HB/jbruner.html

Bruner, J. 1990. *Acts of meaning*. Cambridge, MA: Harvard University Press.

———. 2002. *Making stories: Law, literature, life*. Cambridge, MA: Harvard University Press.

Carter, K. 1993. The place of story in the study of teaching and teacher education. *Educational Research* 61, 157–78.

Chomsky, N. 1965. *Aspects of the theory of syntax*. Cambridge, MA: M.I.T. Press.

Clandinin, D.J., and F.M. Connelly. 2000. *Narrative inquiry: Experience and story in qualitative research*. San Francisco: Jossey-Bass Publishers.

Fivush, R. 2006a. *Memory and narrative, self and voice*. Retrieved January 17, 2006, from http://narrativematters.com/speakers.html

———. 2006b. *Family narratives project*. Retrieved January 17, 2006, from http://psychology.emory.edu/cognition/fivush/lab/Family%20Narrative.htm

Freeman, D. 1989. Teacher training, development, and decision making: a model of teaching and related strategies for language teacher education. *TESOL Quarterly* 23(1), 27–45.

———. 1996a. The "unstudied problem": Research on teacher learning in language teaching. In D. Freeman and J. Richards (Eds.). *Teacher learning in language teaching* (pp. 351–78). New York: Cambridge University Press.

———. 1996b. Redefining the relationship between research and what teachers know. In K. Bailey and D. Nunan (Eds.). *Voices from the language classroom* (pp. 88–115). New York: Cambridge University Press.

Halkes, R., and J. Olsen. 1984. *Teacher thinking: A new perspective on persisting problems in education*. Lisse, Netherlands: Swets and Zeitlinger.

Holliday, A. 1994. *Appropriate methodology and social context*. Glasgow: Cambridge University Press.

Koike, D.A., and J.E. Liskin-Gasparro. 2003. Privilege of the nonnative speaker meets the practical needs of the language teacher. *The Sociolinguistics of Foreign-Language Classrooms: Contributions of the Native, the Near-Native, and the Non-Native Speaker. Issues in Language Program Direction*. Retrieved August 9, 2008, from http://eric.ed.gov/

Kramsch, C. 2003. The privilege of the non-native speaker. *The Sociolinguistics of Foreign-Language Classrooms: Contributions of the Native, the Near-Native, and the Non-Native Speaker. Issues in Language Program Direction*. Retrieved February 2, 2006, from http://eric.ed.gov/

Lightbown, P.M., and N. Spada. 1993. *How languages are learned.* Hong Kong: Oxford University Press.

Long, M. 1980. Inside the "black box": Methodological issues in research on language teaching and learning. *Language Learning* 30, 1–42.

Mattos, A.M.A. 1997. Native and non-native teacher: A matter to think over. *English Teaching Forum* 35(1), 38.

———. 2000. *Percepções de uma professora de inglês sobre sua sala de aula: Uma visão êmica.* M.A. Thesis, Language Institute, Federal University of Minas Gerais.

———. 2002. Percepções sobre a sala de aula de língua estrangeira: Uma visão global e êmica. *Revista de Estudos da Linguagem* 10(1), 109–38.

Medgyes, P. 1994. *The non-native teacher.* Hong Kong: Macmillan.

Pavlenko, A. 2001. Language learning memoirs as a gendered genre. *Applied Linguistics* 2(22), 213–40.

Richards, J. 1998. *Beyond training.* New York: Cambridge University Press.

Richards, J.C., and D. Nunan (Eds.). 1990. *Second language teacher education.* New York: Cambridge University Press.

Stern, H.H. 1983. *Fundamental concepts of language teaching.* Hong Kong: Oxford University Press.

Woods, D. 1996. *Teacher cognition in language teaching: Beliefs, decision-making and classroom practice.* Cambridge: Cambridge University Press.

CHAPTER 3

Becoming a Teacher: Using Narratives to Develop a Professional Stance of Teaching Science

Robert W. Blake Jr. and Sarah Haines

> Once they [preservice teachers] complete teacher education programs, new teachers enter schools where practices engender particular conceptualizations of...teaching. The conceptualizations of...teaching that are promulgated in these different settings are not all consistent. A lack of consistency across contexts can impose stresses on new teachers and render their professional development complex.... How do they manage the contradictions across different conceptualizations? How do they succeed in their contexts of practice as new teachers?
> —Trumbull 1999, p. xv

> ...teacher education is an ongoing process of inquiry in which there is a continuous dialogue between theory and practice.
> —Clandinin, Davies, Hogan, and Kennard 1993, p. 211

Why Narratives of Preservice Science Teachers?

Both Trumbull (1999) and Clandinin et al. (1993) provide statements that act to rationalize and clarify the use of narratives as a means of understanding the teaching process. From practicing teachers and preservice interns narratives can be used as a way to "interrogate" practice to "get at" an understanding of what one knows and does not know (Lyons and Labosky 2002). In this sense narrative is a way of "constructing the knowledge of teaching" (Lyons and Labosky 2002, pp. 21–22), or even reconstructing a belief system of what it means to be successful as a teacher. Over the years we have seen a variety of writings that focus on narratives within the profession of teaching (Blake 2004; Jalongo and Eisenberg 1995; Kooy 2006; Mack-Kirschner 2003; Schubert and Ayers 1992; Trimmer 1997; and Trumbull 1999). These texts have helped to lay the foundation for the use of narratives as a means of understanding practice. For researchers, narratives can be used as a method of inquiry that allows us to better understand the processes of becoming a teacher, and to articulate these stories and lessons learned in our own teaching practices with preservice and in-service teachers.

Stories gleaned from teaching can be a powerful mechanism for conveying and clarifying the intricacies of the acts within the complex social process. The narratives or stories can help to construct an understanding between the ideals of teaching (often the "theories" of teaching as presented in teacher education) and the actual practice of teaching in the realm of a real classroom.

Narratives as Reflection: Bridging Theory and Practice

Bryan and Abell (1999) ask, "How do we help them [student teachers] to articulate, analyze, and refine their beliefs about teaching and learning?" (p. 172). Korthagen and Kessels (1999) propose that teacher educators move beyond an "application-of-theory model" (p. 4) to "reflective approaches" where students are provided opportunities to consider their teaching experiences within the context of actual classroom practice. Bryan and Abell (1999) suggest that incongruent messages from the university and school cultures may cause conflicts that inhibit an intern's ability to reflect on teaching and link, or bridge, their theory and practice. Teacher preparation programs are often accused of focusing primarily on theory, the ideal of teaching where students are simply expected to transfer what they have learned at the university into their classroom experience. The school culture, on the other hand, is thought to center on pragmatic experiences, or the reality of teaching, with little perceived connection to theory. The difficulty for preservice interns, therefore, is that they are expected to incorporate their personal theory of teaching, which may include university ideals into their classroom practice as they struggle to become a teacher.

In light of this issue we use narratives in science education not only to link theory and practice but also to showcase the richness and reality of these experiences that is not apparent in a more formal presentation of scientific data. Such traditional data is representative of a paradigmatic/scientific way of knowing, a mode of thought that seeks objective truth in an empirical, scientific sense (Bruner 1985, 1986). We agree with Bruner that the narrative and paradigmatic ways of knowing are "irreducible modes of cognitive functioning" (1985, p. 97). In using narrative we counter the paradigmatic and seek "truth-likeness," as opposed to objective truth, as a means of "getting at" each intern's sense of teaching science.

By encouraging a reflective process we seek to gain insight into how preservice teachers "manage contradictions across different conceptualizations" of teaching science (see Trumbull 1999) and thus, how they create a link/bridge between a theory of teaching science to a practical set of beliefs that allow for successful classroom teaching.

Teacher Beliefs

As we consider the personal theory as a foundation of teaching we must also consider the relationship between these theories and teacher beliefs. As Pajares (1992, p. 307) puts it:

> Few would argue that the beliefs teachers hold influence their perceptions and judgments, which, in turn, affect their behavior in the classroom, or that understanding the belief structure of teachers and teacher candidates is essential to improving their professional preparation and teaching practices.

Teacher belief research indicates a strong link between the practice of teaching and what teachers believe about teaching. (Clark and Peterson 1986; Kagan 1992; Munby 1982, 1984; Nespor 1987; Tabachnick, Popkewitz, and Zeichner 1979) It seems reasonable to assume, therefore, that beliefs are structured from previous events and experiences and thus, any past event can create guiding images that act as a filter for new information. In addition, while beliefs are resilient and difficult to change they are used as a standard that newer information is compared to and thus, may be important to ascertain early in the teaching career.

The Professional Stance

For each of us the teaching journey begins with our *personal theory* of teaching science, a set of beliefs constructed from the interrelationship of experiences which include classroom situations (elementary, middle, and high school) as well as university coursework. As Trumbull (1999) outlined in her study, preservice teachers may face a clear disconnect between the theoretical frameworks of teaching presented at the university and the practical/pragmatic solutions to teaching advocated in public schools. Preservice teachers may often find that they are expected to incorporate college theory into actual classroom practice. The pressure of theory implementation can create a conflict; a struggle between the practical demands of public schools with the theories presented by university professors. Through this struggle, or interrelationship between theory and practice, each of us may modify our ideas of teaching into a set of beliefs that enable us to teach in the pragmatic, day-to-day aspects of our profession.

We define this revised set of functioning beliefs as the *Professional Stance*. As a new teacher gains experience, the *personal theory* is reconstructed relative to the culture of schooling. The professional stance, whatever it may become, is ultimately the central guiding system of beliefs that influences what the teacher does in the classroom. Conceptually, the professional stance does not replace a personal theory, but is a reconstruction of that theory; a template of beliefs that allows a teacher to teach in the classroom. The stance, therefore, becomes a means to bridge theory and practice, allowing teachers to exam the practical issues as they attempt to link these issues to theoretical aspects of teaching.

Narratives can be used by preservice teachers to gain insight into how they link their original theories of teaching science to actual classroom practice, as well as how they "manage" the contradictions they may face in the day-to-day setting of teaching.

The narratives presented here are an initial glimpse into the beliefs of preservice interns and provide insight not only into what each thinks is important for teaching science, but also provides us with ideas on how we may reconstruct our own process of teaching preservice teachers of science.

Narrative Collection and Analyses

Narratives, as written pieces, pose a particular challenge as a research tool. Many times we tend to fragment these writings into manageable "chunks," ones that we, the researchers, identify. In discussing the use of the text as a whole, Kohler-Riessman (1993, 2008) asserts that making sense and meaning of a particular passage, understanding why a person acts or reacts in a certain way, or just plain listening to a person's story cannot necessarily be segmented and fractured into small, manageable pieces. However, in reviewing narrative passages, Kohler-Riessman (2008) offers four general ways to approach the analysis process. These four methods are:

1. Thematic
2. Structural
3. Dialogic/performance
4. Visual

Here we analyze the passages thematically, based on "told" events, ones that are generated mainly from provided prompts (as opposed to those of "telling"—see Kohler-Riessman 2008, p. 54).

Ideally we prefer stories based on "free writing," but we provided prompts to aid in the initial writing process and to provide an avenue of writing for those who do better with prompts. We collected preservice teacher narratives from weekly writings related to experiences in the secondary science student teaching semester. Table 3.1 outlines the source of these narratives. All interns were asked to write once per week with no stipulations on the length or even the topic if they chose to free-write. They were, however, consistently reminded that the theme was "*becoming a teacher of science*" and that in no way was their grade tied to the responses.

While analysis was ongoing, two general themes emerged from the writing. These were:

- The identification of *enduring beliefs* and practice of teaching science, and
- The identification of characteristics that ensure the intern is *becoming a teacher*.

Table 3.1 Source of Written Narratives

General Prompts

1. What happened this week that helped you to become a science teacher?
2. What happened to hinder your becoming a science teacher?
3. What are you doing yourself to become a science teacher?

Open Ended/Affective Writing

1. Describing experiences and/or incidences
2. Responding to reader comments/questions

Final Prompt

I know I became a science teacher when…

Source: Author's Illustration

The Narratives

For each of these themes we present narratives from two preservice secondary science teachers. Our goal is to look at the "told" events and discern how each intern's thought processes move toward becoming a teacher as each constructs a professional stance. While we realize that making generalized statements of understanding from these short excerpts is incomplete, we do believe that these collections act as a beginning in ascertaining how preservice interns' thoughts of teaching develop into a functional set of beliefs, ones that allow them to be successful in the classroom. In addition, we, as science educators, can use these narratives as a means to reflect on our own practice of how we engage preservice teachers in establishing what they believe to be important in teaching science.

Theme #1: Enduring Beliefs of "Becoming a Teacher"

In this section we present narratives from two secondary science interns: Chelsea and Kim. We used prompts to support their initial narrative process and hoped that they would move from these guided responses to more open-ended narratives; ones that allowed them to explore and expand into topics that were of most interest to them. Chelsea and Kim both wrote narratives each week explaining where they thought they were in the process of becoming a teacher, and both adhered closely to the provided prompts. We begin with Chelsea's narratives and then move directly into Kim's story.

Chelsea, February 11, 2005

What happened to hinder your becoming a science teacher? As I think back on this week and recall what has hindered me becoming a science teacher it would be that science can be a hard subject to explain. After a lesson I taught this week, I was thinking about how I can ask better questions to students and how I am able to best explain an assignment without giving away the answer, and in-general just to explain certain material which seems like it is complex or above students knowledge. I am finding it difficult to know how to tell whether or not material is too difficult for the students. Along with this, how do I know how to explain the material again when they don't get it the first time. This has hindered me because it has made me think that science is a difficult subject to teach.

What are you doing yourself to become a science teacher? To become a science teacher I am trying to strengthen my knowledge of science and to learn ways in which to best describe science to my students using analogies which will relate science, or the topic I am teaching, to students everyday experiences.

Chelsea, February 25, 2005
   What happened this week that helped you to become a science teacher? At the close of this week I feel much more confident about becoming a science teacher than I have since the beginning of my teaching experience. Around the middle of this week we began to introduce a new topic on rocks and minerals. The lesson this week that has really encouraged me in my teaching experience for all of my classes was a split class between notes and a lab/exploration activity. During this lesson students were introduced to minerals through note taking and then they were able to explore the different rocks at the lab stations. I felt this lesson helped me to become a science teacher because I was able to explain a lot of background information on minerals to my students and I was able to engage them in the material and have them be excited about the rocks and minerals they would see at the stations.

   What happened to hinder your becoming a science teacher? As I think back on this week to recall what hindered my becoming a science teacher, I feel it was a lesson we were completing earlier in the week on earthquakes and volcanoes. This lesson was discouraging because I felt I didn't know enough about the activity we were completing to have it be successful. I was not too excited about the activity the students were completing and I feel this hindered my ability to teach the material successfully.

Chelsea, May 16, 2005
   Final Reflection. I knew I became a science teacher when I was able to feel more comfortable being up front and teaching the content to my students. I also knew I became a science teacher when I was able to see my students succeed on the final quiz I gave to them. These two things have helped me to realize I have become a science teacher, because I am able to see that I have been successful in teaching students the necessary knowledge they need to know in order to be successful. This alone helps to give one confidence to keep on going because you know that your students were able to take something away from the class, which you taught them. It is a very rewarding experience to know that you have impacted the life of another person who will go and do great things someday.

Kim, March 28, 2005
   Looking back on myself as a beginning student teacher to where I am now at the end of my first rotation, I can see many areas in which I have grown. I have learned to be confident in myself in front of a class. Projected confidence can do wonders for classroom management. Students will accept and forgive simple mistakes made in class if the teacher is humble about

it. I made many mistakes at the board, but when a student pointed it out, I began to see that acknowledging it and moving on seemed to be the best route. I think my content knowledge still gained the students respect. I also learned that a quickly moving lesson with plenty of activities planned keeps a class on its feet and minimizes behavioral problems. It was the lesson in which I had down time that the misbehavior started to arise. I also realized that a lesson can be too packed. Students bombarded with a heavy list of activities at the beginning of a mod may shut down on information overload or feel rushed. Careful planning with ample activities but a few main goals turned out to be the happy medium. Giving the students a few main activities with plenty smaller activities planned just in case worked out fine. Finally, from a different standpoint, I learned more skills in the time management department. As a student teacher also attending one night class, I learned to manage my time between the two and still give my full attention to both. My mentor teacher commented that this was a good experience because teachers are constantly having to take classes or attend meetings after school and having that extra class gave me a taste of what the teaching career is really like.

Kim, April 5, 2005

Overall, my experience at PHHS was great. I was really pleased with my performance, my mentor, the students, and the school as a whole. The "standard" students were very receptive of me and were sad to see me go. I had a couple of them write me letters. One was a disruptive, energetic, bright, misled young man. I gave him detention for his behavior and during so I gave him a little talk. He wrote me a letter thanking me and saying that he realized what I was saying and that he will try hard and get his stuff straight. The letter was very touching and it meant a lot. I had the students evaluate my performance, and I am glad that I did. For the most part, they were very good and helpful. The "GT" students were very nice but were not as attached to me. I really wish I was able to stay at PHHS.

My goals for the middle school:
* Be more creative with lessons and activities
* Do more activities/labs and LESS lecture
* Integrate students in their own learning: help them understand how to learn and study, etc.

Kim, April 18, 2005

Final Reflection. So far, I feel as if I have worked towards my goals almost everyday. I don't think that I have even "lectured" once. It seems that in middle school, it is next to impossible to teach without fun activities, creativity, group cooperation, and lab work. I feel as I plan every lesson with the kids' attention spans and interests in mind. If they are not focused and engaged they won't learn. It is a whole different world teaching middle school. Some things are much easier (planning) and some things are not (energy levels). Either way, I do enjoy it. I have been very lucky to get the group of students and teachers that I have in this team.

p.s. I knew I was a science teacher when...I was completely lost when I forgot my planning book at home.

Comparing Chelsea's and Kim's Enduring Beliefs (A Snapshot of the Whole Movie)

From the writings of Chelsea and Kim we were able to gain a glimpse into what we considered the *enduring beliefs* of each intern. In reading these pieces it seemed that each had a fundamentally different view on what was important for the teaching of science. Chelsea focused on her confidence and her ability to know and teach the content of science. As former science teachers we fully understand her concerns. If asked, we suspect her teaching metaphor for her role as a teacher would resemble a *dispenser of knowledge*, one whose main goal is to ensure student content understanding. Kim, on the other hand, emphasized pedagogical aspects and her ability to engage students in learning and to provide them with experiences that offer variation of experience all while promoting inquiry. Table 3.2 compares Chelsea's and Kim's enduring beliefs.

From these excerpts what is interesting and potentially applicable for science educators is that both interns took the same classes, heard the "same" messages regarding inquiry and constructivist learning, and yet appear to have interpreted and internalized the course work to fit their individual set of beliefs. This notion of fitting new material with current beliefs is well documented in the teacher belief literature mentioned earlier and is a clear example of how short-term experiences may not impact or even change a person's fundamental beliefs.

What was not done by us at the beginning of the interns' experience was a direct attempt to ascertain a fundamental system of beliefs, an earlier analysis of narratives by both intern and teacher. Doing so may have provided each an opportunity to address, to know, and to understand what it is that she believes regarding the teaching of science. Any new information and experiences could, therefore, be used to address issues and areas that they themselves wish to change. This simple activity would not only have allowed the interns to examine enduring beliefs, but also open possibilities

Table 3.2 Comparing Enduring Beliefs

Chelsea: Confidence in knowing, explaining, and teaching content appear to be the driving force.

- "science can be a hard subject to explain."
- "I felt this lesson helped me to become a science teacher because I was able to explain a lot of background information on minerals to my students and I was able to engage them in the material…"
- "These two things have helped me to realize I have become a science teacher, because I am able to see that I have been successful in teaching students the necessary knowledge they need to know in order to be successful."

Kim: Planning, variation in method, and fast pacing appear to be the driving force

- "fun activities, creativity, group cooperation, and lab work."
- "Careful planning with ample activities but a few main goals turned out to be the happy medium."
- "I also learned that a quickly moving lesson with plenty of activities planned keeps a class on its feet and minimizes behavioral problems."

Source: Author's Illustration

for changing and addressing how we, as teacher educators, attend to the teaching and learning of science for preservice interns.

Theme #2: I Know I Became a Science Teacher When…

The second source of narratives comes directly from the end of semester prompt: *"I know I became a science teacher when…"* This prompt was intended to elicit a response that portrays the moment (the week? the day? the lesson?), within the overall teaching process that each intern can identify as the "tipping point" to *becoming a teacher* and to have sense of a professional stance.

We begin with Becky, a chemistry major, and her depiction of her thoughts on when she became a teacher of science. For Becky, it appears that "ownership" is her key.

Becky

End of Student Teaching Program.

I know I became a science teacher when I moved to the middle school and began discussing how we wanted to do things with my new mentor teacher. I feel like I have somewhat always felt like a teacher, like it was in my blood. I have known all my life that this was the profession for me. However, as we began talking I found myself telling about the daily routines that I like to keep with *my students* and the activities that I like to do with *my students*, etc. It took me a minute to realize that at some point during my high school experience I had started referring to my environment as *my classroom* and *my students*. I didn't realize it until then because I had never been questioned. If I told a student not to pass notes in *my classroom* or was discussing *my classes* with another teacher, no one thought of the situation otherwise. I realized that slowly I had taken over the classroom and made the transition from the student teacher to the teacher without knowing it. It was at the point of my discussion with my middle school mentor teacher that I realized what had happened. I was excited about the fact that I really hadn't noticed and more excited that no one else did either. I feel that as I move into my own classroom with my own styles and setups, I will feel even more like a teacher because it will be up to me to hold my self in that position. I think that it is because I had great mentor teachers that I really didn't have trouble with that transition.

Jennifer's response is also from the final prompt but is within the context of the semester before student teaching. We showcase this particular narrative because we were intrigued by her insightfulness of what she considered the characteristics of a good teacher. While not explicitly stated nor purposefully addressing this topic, Jennifer provides specific examples of what she believes to be the traits of good teaching. In this narrative we have identified those statements by using italics.

Jennifer

"I know I became a teacher when...." Theme:

I'm getting there! I just did another bit teach at the middle school and it went really well. Mrs. S let me teach an introductory lesson on the skeletal system for two of her morning classes (both standard classes). She had given me some material to work from (a segment of a power point presentation, a video, and some worksheets) and said I could use all or none of them. We agreed the Thursday before that I would focus the lesson on skeletal system functions. She said that she liked what I had e-mailed her the Sunday before the lesson (*Planning and Preparation*), so I "tweaked" a few things and went with what I had (*Establishing Ownership*). I ended up using the power point slides, but added a few of my own into the slide show. I used one of the worksheets because it went well with the slides. I also used the video because I felt it was a good introduction to bones, and the bone segment was only 8.5 minutes. (*Generating Variety within the Lesson*) I created a worksheet to go along with the video so the students would have something to do and not fall asleep. (*Establishing Student Accountability*) All of the teachers at the high school said that giving the students something to do during videos was a must! The morning of, Mrs. S said that she could get me a skeleton and asked if I wanted to use that. I told her that would be great. She came back with the skeleton...with no head. She asked if I wanted the skull and I told her I'd go to the storage room to find it. I was only able to find the very top portion of the skull, which is really the only part that I wanted anyway (to show the immovable joints/suture lines), so I took that back to the classroom. "Bob, the headless skeleton" did come in handy and the students really liked him.

I wasn't nervous. (*Establishing Confidence in Content as well as a "Teaching Presence"*) I didn't feel my face turn lobster red (which it does when I'm self conscious). I wasn't fixated on whether I knew the material well enough, because I was comfortable with it...and I actually was able to answer questions about bones that we weren't going to necessarily talk about that day, so I didn't prep/review for it. I didn't talk fast...well, I didn't talk THAT fast! The students did seem to be keeping up, so it wasn't a huge issue. I was able to use a bunch of different activities and tactics to get the point across (movie, worksheets, skeleton, lecture from power point). I also had a student volunteer come up to the front of the class to demonstrate joint movements...the students liked that because the volunteer looked kind of goofy doing all of the movements (fan/Karate kicks, knee jerks, swinging arms around, lifting invisible weights...losing their balance while trying to demonstrate ball and socket hip joint movements)! It was fun!!! (*Variety of Engagements*) We also had a discussion on when bones don't do what they're supposed to...osteoporosis, arthritis, broken bones...and how that affects everyday life. (*Application of Understanding/ Linking to Life*) Some of the girls were a little disgusted (but entertained) by the idea of forcing your elbow joint to move the wrong way.

I always had lots of hands to choose from to answer questions but I made a point to occasionally pick on someone who wasn't raising their hand or if it was their first time raising their hand. (*Generating and Maintaining Student Involvement*) I was better with the casual speaking, but I still need to work on that a bit more. The lesson didn't feel like a performance anymore, it was more of a discussion or dialogue between me and the students. I actually felt like a teacher. It was great!!!! (*Engagement/ Interactions/Generating Relationships*)

After the lesson, I gave the students an exit ticket that asked them to list three of the four functions that we talked about and explain them (i.e. what would happen to you if that function failed, why that function is important, etc.). Most students did a great job, but there were some that really didn't get it. (*Formative Assessment*) At first, I was a little upset because we talked about the functions at least three times during the lesson, but then I realized where some of them may have gotten confused. I used most of the power point presentation as is. What I should have done was create another slide that explicitly listed the four functions (i.e. #1, #2, #3, #4). I think that would have helped. In addition, I mentioned support as a function at the beginning of the lesson (showed a picture of an inflated Santa balloon "body with bones" and another picture of a deflated Santa balloon "body without bones"), but it wasn't one of the functions that Mrs. S really wanted them to know. (She did say that she liked the slides though). She lumped support in with movement and I'm not sure that I made that completely clear. (*Reflective Practice and Readjusting Learning*)

Mrs. S said that I did a great job and so did the middle school's mentor teacher (who observed a portion of the lesson). Mrs. S actually wrote me a note that she said that I could use as feedback if I needed written documentation for anything. In her note, she said "it is easy to spot a true teacher." I guess if someone I've only known for a week thinks that I will be a good teacher, and who herself is a very good teacher, I should probably stop being quite so hard on myself, (*Self-Criticism/Reflective Practice*) relax and have fun. I really am starting to believe that I can do this and be really good at it. I know this is what I'm supposed to be doing because I'm having so much fun with it! It doesn't feel like a job to me and I look forward to being in the schools each day. (*Becoming a Teacher*) I can't wait to start student teaching!! ☺

Thoughts on Becky and Jennifer

While showcasing two excerpts form different timeframes of preservice teaching, Becky and Jennifer do have commonalities in their responses. Becky addresses a sense of *ownership* as the impetus of when she knew she became a teacher, and we believe that the issue of ownership is vital as preservice teachers prepare for the totality of the profession. One difficulty in student teaching is often an intern's lack of owning her teaching. This particularly happens at the beginning of the process where most of the time the classroom, the lessons, and the assessments "belong" to the mentor teacher and school system. For Becky "owning" the entire process of teaching, attaining a sense of possession of planning, implementing, assessing, and even managing discipline seems to have led to a strong sense of becoming a teacher. With increased ownership the teaching process becomes "real," where an intern is able to take full responsibility for most all of the decisions she makes.

In this piece Becky does not talk about the content of science, but speaks of ownership in a more general sense; a more generic attribute in becoming a teacher. Quite possibly, though, if prodded, we suspect that Becky might discuss her sense of ownership of the science content, just as Chelsea did. However, even without a content specific discussion we cannot discount the importance that this ideal of ownership has within the larger notion of becoming a teacher. While Chelsea and Kim both had different enduring beliefs it is these beliefs that most likely lead to "ownership," or a feeling of finally becoming a teacher. Thus, explicitly addressing intern autonomy and ownership may allow them to take control in developing management styles, lessons, and assessments and may greatly impact the relative pace of becoming a teacher.

Jennifer's response was also appealing because it came from the semester before student teaching. Originally, we had not intended to include any narratives from that semester, but we were intrigued by her insightful and detailed description regarding the qualities of a "good" teacher (see table 3.3). Not only is her list useful, again from a holistic standpoint, but we also gain insights as to what Jennifer deems important as she strives to

Table 3.3 Jennifer's Qualities of a "Good" Teacher

- Planning and preparation
- Establishing ownership
- Generating variety within the lesson
- Establishing student accountability
- Establishing confidence in content as well as a "teaching presence"
- Variety of engagements
- Application of understanding/linking to life
- Generating and maintaining student involvement
- Engagement/interactions/generating relationships
- Formative assessment
- Reflective practice and readjusting learning
- Self-criticism/reflective practice

Source: Author's Illustration

become a high-quality science teacher. The next stage would be to follow Jennifer through student teaching, and beyond, to compare/contrast her narratives as she becomes a teacher.

Developing a Professional Stance

Presented here is but one example of how narratives can be used to gain an understanding of the developing belief structures of preservice teachers. As we continue to ask these interns to reflect and write on their experiences of becoming a teacher, we persistently view these stories through the *theory-practice* lens; a lens that attempts to focus on the process of becoming a teacher as each constructs a *professional stance*. Ideally, our purposes are twofold. First, we use narratives as a means for interns to address their enduring and changing beliefs of what it means to teach science. We provide each person opportunities to question their own thoughts of teaching as these ideas are not only juxtaposed with the ideas presented in university coursework but also with the practices presented in the actual classroom experience. Thus, we have a *theory-practice* association. We wish to not only help student teachers manage the contradictions but we also wish to gain insight into how they do this.

Second, we hope to use the stories to address our own methods and conceptions of teaching preservice science interns. We wish to gain a clearer understanding of the issues that these teachers deem most important in becoming a teacher so we can adjust our own practices to meet these needs. In doing so we may actually reconstruct our own professional stance.

We also believe that using narratives from past interns as a learning tool for future science teachers could provide incoming interns with stories that reflect problems and issues regarding the teaching of science. We believe that this personalization of the teaching process will provide interns an opportunity to more critically analyze their own beliefs and views of what it means to become a teacher. Similar to case study analysis, narratives can be used to garner an understanding about specific and general teaching experiences. We believe, however, that narratives take this process further where the narrator not only presents an issue or problem but also proposes a solution, or at least shares experiences from decisions made to address the issue. It is the narrator's discussion of these potential solutions that allows the audience a view into "real" stories and to compare these to their own ideas and experiences. Thus, the power of narrative is in the reality of the story, the "truthlikeness" (Bruner 1985, p. 97) and the ability of story to connect meaning, either affirming or disaffirming the reality of one's own experience.

References

Blake, R.W., Jr. 2004. *An enactment of science: A dynamic balance among curriculum, context, and teacher beliefs.* New York: Peter Lang.

Bruner, J. 1985. Narrative and paradigmatic modes of thought. In E. Eisner (Ed.). *Learning and teaching the ways of knowing. Eighty-fourth yearbook of the national society for the study of education. Part II* (pp. 97–115). Chicago: University of Chicago Press.

———. 1986. *Actual minds, possible worlds.* Cambridge, MA: Harvard University Press.

Bryan, L.A., and S.K. Abell. 1999. Development of professional knowledge in learning to teach elementary science. *Journal of Research in Science Teaching* 36(2), 121–39.

Clandinin, D.J., A. Davies, P. Hogan, and B. Kennard. 1993. *Learning to teach. Teaching to learn.* New York: Teachers College Press.

Clark, C.M., and P.L. Peterson. 1986. Teachers' thought processes. In M. Wittrock (Ed.). *Handbook of research on teaching.* 3rd ed. New York: Macmillan.

Jalongo, M.R., and J. Eisenberg. 1995. *Teachers' stories: From personal narrative to professional insight.* Hoboken, NJ: Wiley, John & Sons.

Kagan, D.M. 1992. Implications of research on teacher belief. *Educational Psychologist* 27(1), 65–90.

Kohler-Riessman, C. 1993. *Narrative analysis. Qualitative research methods series* 30. Thousand Oaks, CA: SAGE Publications.

———. 2008. *Narrative methods for the human sciences.* Thousand Oaks, CA: SAGE Publications.

Kooy, M. 2006. The telling stories of novice teachers: Constructing teacher knowledge in book clubs. *Teaching and Teacher Education* 22(6), 661–74.

Korthagen, F.J., and J.P. Kessels. 1999. Linking theory and practice: Changing the pedagogy of teacher education. *Educational Researcher* 28(4), 4–17.

Lyons, N., and V.K. Labosky. 2002. *Narrative inquiry in practice: Advancing the knowledge of teaching.* New York: Teachers College Press.

Mack-Kirschner, A. 2003. *Powerful classroom stories from accomplished teachers: Stories from the classrooms of accomplished teachers.* Thousand Oaks, CA: SAGE Publications.

Munby, H. 1982. The place of teachers' beliefs in research on teacher thinking and decision making, and an alternative methodology. *Instructional Science* 11, 201–25.

———. 1984. A qualitative approach to the study of a teacher's beliefs. *Journal of Research in Science Teaching* 21(1), 27–38.

Nespor, J. 1987. The role of beliefs in the practice of teaching. *Journal of Curriculum Studies* 19, 317–28.

Pajares, M.F. 1992. Teachers' beliefs and educational research: Cleaning up a messy construct. *Review of Educational Research* 62(3), 307–32.

Schubert, W.H., and W.C. Ayers. (Eds.). 1992. *Teacher lore: Learning from our own experience.* New York: Longman.

Tabachnick, B.R., T.S. Popkewitz, and K.M. Zeichner. 1979. Teacher education and the professional perspectives of student teachers. *Interchange* 10(4), 12–29.

Trimmer, J. (Ed.). 1997. *Narration as knowledge: Tales of the teaching life* (pp. ix–xv). Portsmouth, NH: Boynton Cook.

Trumbull, D.J. 1999. *The new science teacher.* New York: Teachers College Press.

CHAPTER 4

Personal Politics and Identity in Student Teachers' Stories of Learning to Teach

Alan Ovens

How do student teachers' personal politics affect their development as teachers? In this chapter I examine the principled positions of two student teachers as they tell stories about learning to teach physical education in a New Zealand teacher education program. For one of the students, Vincent, a consistent theme is his identification with Maori (the indigenous people of New Zealand) and his desire to work with young Maori to help improve their life chances. By contrast, the motivation for the other student, Vanessa, is to share the enjoyment of physical activity with people and that everyone can be active as part of their lifestyle. These themes are present in the stories they tell about their journey into teaching and in their connection with aspects of the course. While it is possible that their stories show how they represent formations of identity, my aim is to consider how these principled positions represent political subjectivities that mediate each student teacher's lived experience of learning to teach.

This chapter focuses on the intersection of narratives, memory, and identity. It draws on poststructuralist theories that view identity as constructed in and through narrative. Such theories assume that identity is a fluid concept, highly contextualized, and reflexive (Giddens 1991; Lawler 2008). Rather than reflecting some identity independent of the story being told, identity is seen as being produced in and through narrative as "a set of reifying, significant, endorsable stories about a person" (Sfard and Prusak 2005, p. 14). From this perspective, identity is discursive since it is produced in the process of making sense of various memories, experiences, and interpretations within a coherent narrative of the self. It is reflexive in that it involves the individual continuously monitoring and adjusting to the social contexts and conventions of a given situation (Giddens 1991).

Reflexivity is not always a conscious act. Giddens (1991) suggests that the monitoring of action is integral to what he calls "practical consciousness" (p. 36) that functions "non-consciously" (as opposed to un-consciously) in the continuity of everyday activities. Such practical consciousness allows individuals to act in social situations that require little conscious effort. Rossi and Cassidy (1999) suggest that much of the practice of teaching physical education occurs at the level of practical consciousness and becomes an essential feature of what it means to be and act like a teacher. Reflexivity is also an essential feature of the way students navigate the situations they encounter in teacher education. In the ongoing production of identity, the individual must continually integrate events in the external world and sort them into an ongoing personal narrative. In this way, the political affinities or value orientations of the individual mediate how they approach and make sense of pedagogical discourse. Reflexivity is integral to how student teachers construct situations into meaningful contexts (Clarke and Helme 1993) and determines what it means "to know" (Giddens 1991; Rossi and Cassidy 1999).

A key aim of the analysis is on identifying the subject positions each participant constructs within their stories. I argue that each position can be read as a political position since each prescribes a purpose for education and influences subsequent decisions about curriculum and pedagogy. As a generative theme for thinking about the differences between the participants personal politics, I use Giddens' (1991) distinction between emancipatory politics and life politics.

According to Giddens, an emancipatory politics is concerned with liberating groups or individuals from the constraints that limit their life chances. Power is seen to be hierarchical and divisive and underpinning the exploitation, oppression, and inequality experienced by people. An emancipatory politics makes imperative the ethics of justice, equality, and participation. Emancipatory politics is a politics of life chances. Life politics is not an opposite position to emancipatory politics nor simply a politics that focuses on what happens once emancipation has been achieved. Rather, it is a politics of how one should live their life given the plurality of options deriving from living in a post-traditional world. In this sense, life politics is a politics of life style. Power is viewed as being generative and transformational rather than hierarchical. Giddens (1991) defines life politics in the following way:

> Life politics concerns political issues which flow from processes of self actualization in post-traditional contexts where globalizing influences intrude deeply into the reflexive project of the self, and conversely, where processes of self realization influence global strategies. (p. 214)

Bennett (1998) suggests that because personal identity is now replacing collective identity as the basis for contemporary political engagement, the character of politics is, itself, changing. As a result, there is increasing attention given to managing and expressing complex identities in a fragmenting society. Consequently, political attitudes and actions are less likely to be devoted to grand political projects of social justice and nation building and more toward how best to promote personal lifestyle agendas of how we "ought to live our lives" while being conscious of the consequences of those choices.

The Collective Memory-Work Method

The stories analyzed in this chapter used the Collective Memory-work method initially developed by Haug (1987) and later modified by Crawford, Kippax, Onyx, Gault, and Benton (1990, 1992). Memory-work was selected because of its focus on interpreting participants' subjective experiences through an iterative process of individual and collective analysis of participants' written memories. In general, memory-work involves participants writing narratives about recalled experiences that are then analyzed within the collective research group. The aim, through discussion and reflection, is to achieve an intersubjective understanding of the participants' experiences as the basis for interpreting the research material (Markula and Friend 2005).

Memory-work initially proceeds through a four-step cyclical process. Step one involves each participant recalling a specific memory relevant to the research topic. This is typically done by using a particular phrase or word to help trigger each memory. The four triggering terms selected for this study were "generative spaces," "agency and change," "empowerment," and "personal pedagogy." I chose these terms prior to the study rather than allowing them to emerge from discussion with the participants. My concern was to find trigger terms that would stimulate memories of situations when the student was involved in reflective activity while not predetermining what that reflective activity could be. Since the discrete descriptive memory is central to the method, I also felt I should meet with each participant at the start of each cycle to use a range of strategies, such as drawing, questionnaires, and interviews, to facilitate the identification and recall of a particular memory for each participant to write about.

Following this meeting, step two of the method involved each participant writing a memory-story according to prescribed guidelines (Crawford et al. 1992). Such a process enabled a detailed and descriptive account of the memory to be recorded in a way that enabled the writer to distance themselves from the experience so they could see it anew and not be drawn into interpretation, explanation, or biography (Onyx and Small 2001). The third step of the process involved the group meeting to engage in a collective analysis of each story. Essentially, this was a reflexive process that involved each story being read, discussed, and reflected on. The aim was to allow participants to make sense of their experiences by understanding the social contexts in which their behavior occurred and how they give meanings to the events and actions in their lives (Friend and Thompson 2003). Following this, the fourth step involved offering each participant the opportunity to rewrite their memory-story, modifying it in light of the group discussion.

The final phase involved me reanalyzing all the empirical materials (including those generated as part of the group discussion) in an attempt to "collectivize" the experiences of the students and foreground the key embedded themes. This meant I had to "fracture" the original texts and work with them recursively in a generative manner together with ideas presented in the wider theoretical literature (Alvesson and Skoldberg 2000). In this way, theory is used as one way of making sense of the empirical material rather than the empirical material being used to verify theory.

Vincent and Vanessa were two of five student teachers in their final year of a Physical Education Teacher Education degree who volunteered to be participants in the project. This provided a group small enough to allow for each participant to be involved in the collective discussion and a level of homogeneity that allowed for group compatibility (Kruegar 2000; Small 2000). I then facilitated the group discussions and took responsibility for transcribing tapes and conducting further analysis. Crawford et al. (1992) observed from their experience that composing the research group with the researcher as facilitator rather than as coparticipant reduces the sense of collectivity within the group. In one sense, this could be observed in the way the students limited their participation to writing and group discussion. There was a resistance to full involvement in data analysis and report writing. However, the group was compatible and the level of disclosure, conflict, and discussion was sufficient to provide a productive context for the research to be productive. For the purposes of this chapter, their names have been changed to preserve confidentiality.

The following discussion draws on examples from the stories written by Vincent and Vanessa. Initially, the aim is to show that the subjective positions constructed by each participant represent fundamentally different projects. Following this, the discussion turns to considering how these different subjectivities mediate each participant's lived experience of learning to teach.

Political Projects

Vincent is quite overt about his reason for wanting to teach. His principled position is oriented by a desire to work in some capacity for Maori and, by improving their educational opportunities, improving the quality and conditions of their lives. While this is apparent in many of the comments he makes, it is also evident in the topic, plot, characterization, metaphor, and language choices he makes in constructing both his narratives and comments in discussions. An example of this can be seen in this short extract from a story where he has chosen to write about watching the movie *Once Were Warriors*.

> Yeah, and when I refer to Once Were Warriors it was like when we were in London watching it we were really proud because it was like a really popular movie and afterwards you walk out and way that was a really cool New Zealand movie and stuff. Standing on the marae and seeing everybody and saying yeah, this reminds me of that and it's actually a little bit sad, because that movie was honest I guess, and it was then I started wondering, there must be something else, there must be something better. I've always been brought up in that sort of environment with a little bit of violence I suppose and a lot of drinking. When you're in it it's ok and you become used to it. You know, you hear about what's-his-name had a fight with his missus the other night, but it's not till you get out of it and look from the outside.

Three themes are evident in this extract. First is his identification as Maori. His connection with the movie, feeling proud, standing on a Marae, reflecting on the social environment and culture, all semantically construct his identification with Maori. Second, he uses the movie as a lens to foreground the social problems intrinsic to Maori culture. He suggests the movie was "honest," implying it was in some way telling a truth or revealing something that actually exists. He believes this because it resonates with his own experiences of growing up. He identifies the key problems as being violence, especially domestic violence and the high consumption of alcohol. Third, his comment "there must be something better," when put into context with the other two themes, is the point he reveals his emancipatory politics. His orientation is one of moving away from this position, a desire to be free from the conditions that are creating the problems he is observing.

These themes were also evident in other stories he wrote. For example, in another story Vincent chose to write about joining a program for working with young Maori offenders who ran out of an air force base. The plot focuses on young Maori boys who have criminal records and are in need of guidance and support to better their lives. Within the story he constructs himself as someone with a warmth and sensitivity to these young men and shows a care and concern for wanting to do something to better their lives.

What is highlighted in these two examples is the way Vincent constructs narratives that foreground his identification with and concern for the social problems experienced by young Maori. There is a consistency and coherence in the stories he tells about himself. I would also argue that the subject position he constructs is coherent with the themes of an emancipatory politics.

By contrast, while Vanessa's narratives represents her as also being concerned about issues of fairness and equity, the way this ethic is mobilized is more coherent with a life politics. Vanessa's motivation isn't so much one of working to help improve the life chances of those she is teaching, but more toward individual self-actualization. Her aim is to ensure that physical activity becomes a positive lifestyle choice for individuals and, through her teaching, she can help students make lifestyle decisions. For example, she stated in one group meeting:

> I kind of see it like everyone is put into situations in life and it's up to them to make the most of the opportunity and that teachers probably provide you with those opportunities. (Meeting one, Personal politics and values)

Implicit in this statement is the belief that individuals have both the capacity to make decisions and the agency to act on that choice. Such a belief assumes that power is not hierarchical and oppressive, but generative in the sense that options are created and that responsibility lies with the individual to "make the most of the opportunity." Such a view is coherent with a life politics theme that suggests life is a process of optimal self-actualization from within the plurality of options created by modern society (Giddens 1991).

This theme is also evident in the way she writes her stories. For example, in her story about her personal politics and values she writes about her first day in the hostel as she began her university course:

> Vanessa looked out her window and cringed as the view was packed with buildings and concrete. Vanessa missed everyone already but thought to

herself that many others had been in this same situation, it couldn't be that bad. She was used to making things work for herself and sticking at them. She had saved up enough money for her first year at teachers' college through jobs and scholarships and was not going to pack up simply because she didn't like the look of the place. Vanessa crossed the corridor and looked out the bathroom window feeling relieved to see a huge park through the gaps between the stained hospital chimneys. Vanessa went back to her luggage and pulled out her running gear. She wanted to get outside for a while and take a look around the area that she would be living in for her first year away from home. Home seemed very far away.

In contrast to Vincent's story, Vanessa's is focused much more on herself and contains different themes. She implies she has choices and that she has learnt to make the most of the opportunities life provides. Her attitude is to tap into the generative possibilities provided by the situation. For Vanessa, the opportunity of coming into the university program represents a conscious decision and a sense of self-actualization based on the fulfillment of taking responsibility for oneself financially.

Her extract also contains a theme where the environment plays a role in the plot. She "cringes" at the view of the concrete and buildings and finds solace in the park she can see from her bathroom window. She is someone who enjoys being in the outdoors and having the freedom to enjoy the bush, beach, and open spaces. Like Vincent, it has taken her to move away from what she has taken for granted in order for her to gain an appreciation of this aspect. Caring for and valuing the environment becomes an important theme for Vanessa in various comments she makes and, as Giddens (1991) notes, is also a key substantive moral issue within life politics.

The Reflexive Ordering of Identity

The purpose in the preceding discussion is not to simply label each participant as either exhibiting an emancipatory or life politics. As Giddens (1991) notes, there is no simple dichotomy between the two and in practice the concept is difficult to operationalize. It could be argued that Vincent is himself exercising a life politics in as much as he has exhibited liberty and freedom by being well educated and having the opportunity to travel overseas and, now faced with a plurality of options of what to do with his future, has decided to actualize his potential in helping Maori. Rather, the purpose is to tap into the generative potential of the Giddens' dichotomy to highlight how each participant is motivated by different political commitments and, further to this, consider how such political projects mediate each participant's lived experience.

As argued, Vincent's ongoing story of self is oriented around the themes of identifying strongly with Maori and wanting to help young Maori succeed in the education system. This political project then appears to orient how he makes sense of his own experiences and his perception of the information he is learning. For example, when asked to identify the courses he finds generative for his development as a teacher, he crudely divides them into those that deal with themes of social justice and social transformation, which he labels as "challenging," and those that deal with content, which he feels has little value and labels as "requirements."

His political project also reflexively shapes what he sees as the nature and purpose of physical education. For Vincent, physical education should provide a context where individuals can feel comfortable and gain a sense of achievement. It becomes a location where the focus is not so much on what they are learning, but can instead be a situation where they can connect with someone who can value them as individuals and provide a steadying influence in their lives. It is possible to read this into his story from a practicum experience. In this story he is driving to school and reflecting on his day ahead.

> He went over in his head the three lessons he was teaching for the day. Two in the morning and one after lunch. The morning classes he knew were going to be fun while the afternoon class was always a handful but he loved the challenge. 4D5, a drama class, were a good bunch of kids, at times a little ill disciplined but full of so much vigor and enthusiasm, he knew they would get stuck into whatever he had planned. Seventh Form Bursary was always good. A small class of twelve students, consisting of 6 Maori, 5 Polynesian and 1 South African girl who seemed to know a lot more about Bursary PE theory than any other of the students. As for 3 NO, well you just had to love them, and you had to otherwise they would just drive you insane. The mix was quite refreshing for Jason because he was used to teaching in classes with a higher ratio of European students.

> Vincent reflected on the last three weeks and how he had enjoyed teaching at Pukana College. He thought of the classes he had taught, how he interacted with the staff and students, and in particular what he had gained this teaching experience. He had made a quiet promise to himself that he would try some of the strategies and styles of teaching that he had experienced and learnt at training college. He was in his final year of training, and though in four years he had absorbed a lot of information

about teaching and physical education, a lot of it he felt was irrelevant in his eyes.

Within this extract Vincent contrasts his own pedagogical thinking with the information presented in the program. What is foregrounded in the first paragraph are not just the descriptions of the pupils he is going to teach, but also the key pedagogical issues as he constructs them and his own empathy with the students. He uses terms like "a handful," "at times a little ill disciplined," and "they would drive you insane" to not only imply that behavior is the key issue when working with these students, but also to soften and ease the way this behavior is interpreted. He grounds the students misbehavior in the students own "vigor and enthusiasm" and sees them essentially as a "good bunch of kids." The content of what he is going to teach is not important and is not described. Rather, he uses terms like "fun," "challenge" and "getting stuck in" as descriptors of how he is measuring his lesson.

The first paragraph contrasts with the second in as much as it dismisses the pedagogical knowledge he has been learning within the courses in his degree. In suggesting that he "made a quiet promise to try some of the styles and strategies" he is implying that he views using the styles and strategies is optional. This knowledge is not mobilized to solve the pedagogical problems as he perceives them in the lesson, but rather because they are connected distally to course requirements. This resistance to coursework, however, isn't solely based on an inability to transfer knowledge from one context to another or a lack of appreciation of pedagogical issues. As his various stories demonstrate, he is problematizing pedagogy and questions what pedagogy is appropriate for the type of students he wants to teach. His resistance is directly linked to his reflexive ordering of the nature of teaching that leads to a devaluing and dismissal of the course knowledge as relevant to solving pedagogical issues as he perceives them.

His suggestion in the final sentence of the extract above, that he has absorbed information over the past four years, is also revealing since the absorption metaphor implies that the learning of this information has not been an active process. He appears to have learnt this information almost solely because he has been immersed within it. The mediating factor in limiting his agency has been his perception that it was "irrelevant in his eyes."

Like Vincent, Vanessa values becoming more critically oriented. For example, she identifies on her map of learning to teach the development of "critical thinking" as one of the four factors that have been most generative in developing her ability as a teacher. Like Vincent, she suggests the starting point for this was a course in the first year that focused on the sociocultural foundations of physical education and valued those courses that problematized the knowledge being taught. However, Vanessa's concept of "critical" is fundamentally different to Vincent's. While Vincent is influenced by discourses of social justice and transformation, Vanessa values a critical orientation because of its ability to humanize the process of teaching. For Vanessa, being critical is linked to appreciating the dynamic and complex nature of teaching.

It is evident from her different stories that, for Vanessa, social problems are not conceived as emerging from a hierarchical notion of power or unequal distribution of resources, but are connected with issues emerging from exercising choice within a plurality of lifestyle options. Something of Vanessa's concept of teachers and teaching can be interpreted from comments she made in a group discussion when she tried to clarify a statement she made in a story about seeing meaningless teaching in schools. She comments,

> ...you know how you've got like sport after sport being trudged through and there's no development of things like...you know how people say they want to teach stuff through it. They make use of teachable moments but there is nothing, like there is nothing that is actually set that's planned apart. In my program for third form I had at the beginning of the year a thing where they just become confident in different settings and things like that. Because when they start the secondary school year it it's a whole new environment and just things like that, and then all the units built following different things, but because I felt at other schools if you just do winter sports and a block of summer sports or whatever, if you don't have something more powerful behind that following through, like I don't think I would be motivated to teach it, like I have to have something behind there driving me to think, ok, this is what I'm trying...
>
> ...This is a big generalization saying that I don't like what I've seen out in schools, it's just some schools I've been to. And that's what I was saying at the end. Like for my example here of summer sports. I use the terms summer sports in my thing, but the way I would teach that is whole lot different to the way someone else would teach it and that's the thing, we should have presented in this assignment.

Within her comments Vanessa constructs two images of teaching, that which she has experienced and her ideal. In her experience, the units of work taught by teachers do not build on each other and work to some coherent plan. She uses the metaphor "sport after sport being trudged through" to imply how the units of work she has experienced are being taught with no apparent rationale underpinning their choice or real meaning for the students. The metaphor "trudged" implies a level of persistence and tiredness. To "trudge" through units of work implies that there is no pleasure in the process, the key aim is to get through to the end point. Vanessa's alternative ideal has a strong theme of purpose and coherence to it. Each unit of work builds on the one before and cumulatively works to operationalize her philosophy. Her conception appears to be that teaching is a purposeful activity underpinned by a strong philosophy.

In the way Vanessa conceptualizes teaching, primacy is given to individual agency and missing is any recognition of constraint. It is implied that the teachers should and can implement their philosophy. There is uncertainty if teachers she has seen have a philosophy and a suggestion that they have lost the will to make a difference. Vanessa's perception of poor teaching is conceptualized as an individual problem of energy and effort. There is little recognition of the broader social, economic, or political factors that create work place conditions that constrain teacher agency or lead to fatigue.

Conclusion

This chapter began by asking how student teachers' personal politics influence what and how they learn to teach. By focusing on the stories of two student teachers, I have shown that both use principled subject positions as resources in the narrative construction of their identities. By using Giddens' (1991) differentiation between emancipatory and lifestyle politics it was possible to map some of the differences between their principled subject positions. I have argued that Vincent is motivated by an emancipatory politics and this has influenced not only his desire to be a teacher but also shapes how he perceives the nature of physical education. He is driven by a desire to improve the life chances of the young Maori youth he will teach. Conversely, Vanessa is motivated by a life politics and consequently views the potential of education for its transformative value. For her, the educative potential of schooling lies in its ability for self-actualization of life style choices.

However, the aim was not to explicate and name the political positions of each student teacher. If that were the case a more robust methodology would be required. Rather, the aim was to consider how an individual's personal politics is a reflexive part of their identity that mediates their subjectivity and experience of learning to teach. In this respect it appears that the principled position of each student acts as a significant factor that orients how students make sense of the situations they find themselves in. Typically, a student teacher's biography has been seen as influential in shaping their development as a teacher. The value of this study is to foreground one aspect of this biography and suggest that a greater focus on how the principled positions and value orientations of student teachers will enable a deeper understanding of the teacher education process.

References

Alvesson, M., and K. Skoldberg. 2000. *Reflexive methodology: New vistas for qualitative methodology.* London: SAGE Publications.

Bennett, W. 1998. The uncivic culture: Communication, identity and the rise of lifestyle politics. *PS Online.* Retrieved May 17, 2003, from http://www.apsanet.org/PS/dec98/bennett.cfm

Clarke, D., and S. Helme. 1993. *Context as construct.* Paper presented at the 16th Annual Conference of the Mathematics Education Research group of Australasia, Brisbane, Queensland, July 9–13.

Crawford, J., S. Kippax, U. Onyx, U. Gault, and P. Benton. 1990. Women theorising their experiences of anger: A study using memory work. *Australian Psychologist* 25(3), 333–50.

———. 1992. *Emotion and gender: Constructing meaning from memory.* London: SAGE Publications.

Friend, L., and S. Thompson. 2003. Identity, ethnicity, and gender: Using narratives to understand their meaning in retail shopping encounters. *Consumption, Markets and Culture* 6(1), 23–41.

Giddens, A. 1991. *Modernity and self-identity: Self and society in the late modern age.* Cambridge: Polity Press.

Haug, F. 1987. *Female sexualization: A collective work of memory.* London: Verso.

Kruegar, R. 2000. *Focus groups: A practical guide for applied research.* Thousand Oaks, CA.: SAGE Publications.

Lawler, S. 2008. *Identity: Sociological perspectives.* Cambridge: Polity Press.

Markula, P., and L. Friend. 2005. Remember when...Memory work as an interpretive methodology for sport management. *Journal of Sport Management* 19, 442–63.

Onyx, J., and J. Small. 2001. Memory-work: The method. *Qualitative Inquiry* 7(6), 773–86.

Rossi, T., and T. Cassidy. 1999. Knowledgeable teachers in physical education: A view of teachers' knowledge. In C. Hardy and M. Mawer (Eds.). *Learning and teaching in physical education* (pp. 188–202). London: Falmer Press.

Sfard, A., and A. Prusak. 2005. Telling identities: In search of an analytic tool for investigating learning. *Educational Researcher* 34(4), 14–22.

Small, J. 2000. *Researching different age groups through memory-work.* Paper presented at the Memory-work Research Conference, University of Technology, Sydney.

CHAPTER 5

Learning to Teach Across Cultural Boundaries

Neil Hooley and Maureen Ryan

Introduction

This chapter traces the complicated and emotional journey of two non-Indigenous teacher educators as they work with preservice teachers who are teaching Australian Indigenous[1] children in regular schools. Comprising only a small proportion of the population, Indigenous peoples in Australia are often disadvantaged by the schooling system, being alienated by the non-Indigenous knowledge and approaches to teaching and learning that dominates the curriculum. There are very few fully qualified Indigenous teachers in schools, which is a key factor in making liaison between local communities and the teaching staff very difficult. The story outlines how the authors over time attempt to link major issues in the literature regarding Indigenous ways of knowing with the ideas and concepts of privileged non-Indigenous knowledge in the traditional curriculum. It documents the problems, mistakes, and successes that occur. It takes the notion of two-way learning that is familiar in Australia and links this with the inquiry theories of Dewey (1963), to create the groundwork for a "two-way inquiry learning" curriculum, unique in Australian education. To provide the impetus for this approach in schools, the authors draw on the work of Bruner (1996) and Clandinin and Connelly (2000) to design a narrative curriculum as the framework for two-way inquiry learning. This is a rich, personalized account of a struggle within teacher education to respect and recognize the knowledge and cultural practices of Indigenous communities so that both can form the basis of cross-cultural understanding and teaching in the regular school.

Taking the First Steps

It was most unusual to be in a meeting with a number of Australian Indigenous people being present as well. Our university was receiving a delegation[2] from a country town about two hundred kilometers away from the capital city, Melbourne. As we discussed the possibility of establishing a degree program for Indigenous students to be taught locally, we wondered at the significance of the occasion. Many non-Indigenous Australians have never met an Australian Indigenous person, let alone lived and worked together on a daily basis. This may be due to the fact that less than 3 percent of the overall population are Indigenous and are not spread evenly throughout all aspects of civic life. About 80 percent live in the major cities and regional towns, whereas less than 20 percent live in very small remote communities dotted across the Australian outback. There are major problems regarding Indigenous health, education, housing, and unemployment that remain to be resolved.

The meeting reached agreement on the provision of a four-year Bachelor of Education program for beginning teachers. It would be open to Indigenous and non-Indigenous applicants alike and would respect the culture and knowledge of the community. In general, Australia has very few qualified Indigenous teachers making the linkages between school and community life very difficult. While most Indigenous children attend the neighborhood school and work within the regular curriculum, many do not succeed at the secondary level (Australia 2007). This is probably due to a combination of the social conditions that exist and the nature of the secondary curriculum. Primary schools generally follow a more holistic approach toward knowledge and teaching, whereas secondary schools tend to separate subjects into quite discrete areas. Indigenous ways of knowing (Hughes, More, and Williams 2004) are more congruent with the integrated knowledge of primary schools that usually encourages active discussion of ideas as well.

Our enthusiasm for the task probably masked the difficult road ahead. We knew that there was a gap of about seventeen years between the life expectancy of Indigenous and non-Indigenous peoples across Australia and that the literacy and numeracy outcomes for Indigenous children were considerably lower than for other Australians (OECD 2007). Some progress in "closing the gap" had been made over recent years, but education was still a long way behind. Enrolment of Indigenous students had increased at

Australian universities throughout the 1990s but was now in decline including those in teacher education. Regardless of that, a planning committee was set up and we commenced designing the program to meet university requirements. Amazingly, we completed all the necessary documentation and subject outlines in about eight months, surely a record for university approval processes. After a number of visits and meetings at the town, a small building was leased, furniture and equipment purchased, and teaching began. If any group could make a difference regarding Indigenous education in Australia, we could.

Passion and Principle into Practice

Perhaps some of the best advice we received in the early days of the program was to try to learn *with* the local community rather than attempt to teach *about* the local community. In other words, we should reflect on the issues and problems we were confronting in terms of our own understandings and not attempt to impose what we took as correct knowledge and procedures. We saw three approaches as possible when considering the organization of formal education for Koori[3] communities: cultural and epistemological questions can be ignored, studied in separate subjects, or integrated within all studies. Clearly, the first option is racist and unacceptable. The second option is a common resolution of the issue, but can perpetuate a dichotomy and differential of respect and power between Koori and non-Koori peoples. The third option is preferred because it offers the potential of enriching mutual knowledges and practices and of reconciling a range of antagonisms.

Over a period of time, the planning committee for the program agreed on a number of guiding principles for program design. These involved community responsiveness and partnership, inquiry-based learning, flexible pathways to ongoing education, and development of innovative program practices. The emphasis was placed on holistic knowing in which learners bring their own experiences, history, and culture to bear on specific issues and projects as they are encountered. Our reading at the time indicated a crystal-clear view held by Indigenous communities around the world of what is required for quality Indigenous education (IWGIA 2008). This included highlighting belief and being rather than proof and doing, relationships between broad phenomena rather than disconnectedness, community and society rather than individualism, and what exists in the cyclic present rather than linear arrows of time. In one sense, these characteristics challenge non-Indigenous scientific thinking and have significant implications for schooling and education. In another sense, they are very similar to the notions of integrated knowledge and inquiry learning as advocated by Dewey (1963). Such scientific views may continue the division between ontology, where certain features such as morality are taken as a given for humanness and epistemology, where human modes of thought and belief are constructed through interaction with the universe.

The structure of the Bachelor of Education that emerged from the above thinking was unique in Australian teacher education. It involved two sequences of Koori culture and knowledge subjects over the first three years that were taken by all candidates whether Koori or non-Koori. This is a significant feature given that debate still continues around Australia as to whether all teacher candidates should undertake at least one compulsory unit in Indigenous studies. Other subjects taken within the first three years involved those related to the two curriculum areas of sport and recreation and youth studies. The fourth and final year of the program focused on schooling and curriculum design matters as well as consolidating a reflective approach toward change and improvement. Spread across the four years were 140 days of classroom experience in schools, a major feature of pre-service teacher programs offered at the university. This extensive school experience formed the basis of a strong long-term partnership between the university, community, and schools to encourage a theorizing of practice by all participants. In the first year of the program, all but one of the twenty five students were Indigenous, but with each succeeding year, the number of Indigenous students fell. After four years, overall enrolments reached the anticipated number of sixty, although only 16 percent of these were Indigenous. Teaching staff were mainly non-Indigenous, although administrative staff and guest speakers including Elders and other community members were Indigenous.

If the program was to be genuinely respectful of the interests, culture, and knowledge of the Koori community, then some tricky problems regarding teaching and learning needed to be negotiated. Sessions should be integrated across subject areas and be of informal workshop format and community-based as much as possible. Koori knowledge and ways of knowing should permeate all aspects of teaching and with an emphasis on connections with the land, rivers, animals, plants, and climate. It was intended that when considering issues such as science and the geological features of a particular location, Koori explanations would be sought first and be depicted in song, dance, art, and narrative. Alongside these insights would sit other explanation such as pressure, volume, and temperature that help describe major and minor changes in the physical environment over time. Each explanation of practice is given equal credence and respect, each tempering and enhancing the other. Issues related to country are extremely important in establishing

Indigenous identity, as too are those that relate to family and kin; who you are and where you are from are significant aspects of Koori culture and knowledge that cannot be excluded from educational programs.

Koori explanations of the universe often exist at a level of generality and pattern that would be considered of "higher order" in regular science and mathematics. The attributes of and the relationship between a particular escarpment and area of scrub supporting a specific animal population and the relationship between kin, tribes, and mythologies is a high level of abstraction that seems to have bypassed the process of induction. It fits more readily into a critical consciousness and general paradigm of understanding, even a metaphor or morality of existence and how things are. Modern science is often based on a series of particularities such as the structure of matter, at a level of detail that may not be significant for Indigenous thinking. Here again, the particularities and generalities of different cultures complement each other.

Non-Koori educators are faced with the challenge and dual tasks of attempting to understand both the nature of Koori culture itself in urban and remote areas and the connections that can be built with non-Koori experiences, so that dialectical and integrated investigations of reality can proceed with respect. In progressive educational terms, this can be described as democratic inquiry learning, a procedure that seeks to unite process and content, practice and theory, knowing and doing, mythology and what is considered as scientific fact. A starting point for a university course that incorporates Indigenous perspective may therefore involve the identification by the Indigenous community of key questions that can be pursued for learning purposes and be clarified over a period of time by culturally inclusive research and epistemological heuristics that may be appropriate in connecting diverse cultural understandings and practices. These matters proved to be difficult in practice.

Grappling with Cultural Barriers

For the small number of non-Indigenous staff working in the program, a range of complicated problems needed to be faced every day. These mainly involved issues of culture and pedagogy. It is not the responsibility of educational institutions to teach particular Indigenous ceremony, law, and art that will be sacred and not available for inclusion. Other aspects of history, language, country, and community will however be agreed as appropriate for consideration especially through the ongoing participation of Elders and other community members. Culture (Eagleton 2000; Williams 1981) is a broad concept and is usually taken to include the key ideas, values, traditions, and narratives that inform and guide how we live. It is a dynamic rather than static concept with culture changing as the political and economic nature of society changes. In Australia, for example, we often refer to the vast outback, the distinctive flora and fauna of gum trees and kangaroos, as well as our golden beaches as contributing to a set of defining values including a "fair go for everyone." Educational programs need to be sensitive to not infringing on and misinterpreting culture that is private to particular communities, but there are many other aspects of culture that can be explored. Indeed, a cultural framework is essential for all students to guide their learning.

One university staff member usually made the three-hour car trip early every Monday morning. A combined session for all students began the day, followed by a series of learning circles for different year levels. Learning circles were organized such that small year-level groups were sometimes considering issues that had been presented in the morning session, or were sometimes working on projects that had been negotiated between students and lecturer. Groups would work as independently as possible and would rotate throughout the day meeting with staff as appropriate. This arrangement was devised to enable the small number of staff to be engaged with students as often as possible, but also to meet the requirements of organizing learning around the principles of autonomy, inquiry, and discursiveness. While it was always intended that local community members would be employed as mentors, the lack of people who were available made this a weakness of the program. Final year students operated on a slightly different timetable to enable them to have more time in schools throughout the week and to have at least one full day in conversation with staff.

Because the program had been approved under normal university regulations and had to meet requirements for registration and employment as teachers, it was not the intention that learning be Indigenized. The program did not teach Indigenous culture, Indigenous mathematics, Indigenous science, and the like, but attempted to establish a framework that was culturally inclusive, respectful of the worldviews of different cultures present. In Australia, the term "two-way" (Harris 1990) is used to describe this process. Over time, the program developed the notion of "two-way inquiry learning" (Hooley 2002) that included Dewey's idea of inquiry learning as a means of enabling both cultures to come together on projects of mutual concern so that new ideas and ways of proceeding could be agreed on. This approach contributed to learning circle arrangements as being problem-posing and solving-mechanisms that saw culture as being both stable and changing. Under these circumstances, staff were constantly making decisions and professional judgments about how the process was actually working, and no doubt, mistakes were made. Opening up the question of science teaching in the regular curriculum, for example, meant that staff would attempt to link concepts with local knowledge, particularly the environment. There may be stories about the night sky that could be thought about alongside the evidence of telescopes. Both cultural views were not seen as truth to be taught by teachers and accepted by children, but rather ways that people use to understand and explain their universe. We would like to say that the program constantly endeavored to strengthen its cultural base, particularly in relation to learning from the land, but this may not have been the case.

Many cultural problems were encountered by our Indigenous students undertaking the program, involving both their participation at university and when pursuing their classroom experience in schools. Not only did they need to cope with the efforts of staff at two-way inquiry, but they also needed to work within the regulation and procedure of schools that could be sometimes antagonistic. Dividing knowledge into separate compartments, transmitting difficult ideas to children in short periods of time, not linking knowledge to community issues, not being encouraged to have classrooms that are highly communicative, and assessing learning and children in judgmental ways at specified times all tend to contradict Indigenous ways of learning. It is also difficult for Indigenous teachers to enact forms of discipline and punishment that they see as being quite inappropriate for children. This may not occur to the same extent for Indigenous teacher aides who are employed, but they too are often called upon to intervene when children get into trouble, and the task of mediating between school rules and Indigenous ways can be extremely onerous. The terms of employment for Indigenous teachers in Australia can now include a recognition that cultural time is required for participation in community and family life, including, unfortunately, attendance at the many funerals that occur. Given the lack of Indigenous teachers in Australian schools, Indigenous graduates and especially younger graduates find making the transition to regular classrooms sometimes too difficult to bear.

University staff who were involved in the program and were visiting the campus each week were called upon to walk a tightrope between meeting university and community requirements. Both groups agreed that the program had to be of high quality and not be criticized as being inferior to make the graduation of Indigenous students easier than for their non-Indigenous colleagues in other programs. This meant that staff needed a very strong educational background and the capacity to be flexible and innovative in meeting diverse requirements. Extensive background knowledge of course was also necessary if school subjects like mathematics, science, and history were to be constructed in a two-way manner and be strenuously defended in meeting prerequisites and requirements. Maintaining non-Indigenous enrolments was not a problem, especially with regard to mature age students and in particular women returning to study. It was difficult to maintain Indigenous enrolments however given that only a small number of Indigenous students reach the final years of secondary schooling, and with a small population it is not feasible to keep enrolling mature age Indigenous people as well. If the academic program is not constructed to be culturally inclusive then retention rates of Indigenous students can suffer. It was for a combination of these reasons that the program was most unfortunately discontinued after nine years.

Perhaps the most emblematic issue that confronted us in designing and implementing the program can be seen in what is known in Australia as the Dreaming (Hume 2002). In trying to interpret how Indigenous peoples understood the universe, early anthropologists in Australia struggled with language and notions of science, religion, and spirituality. Of course, early writings in the mid- to late-1800s took place when there were many changes occurring in science, social science, and psychology around the world, so it is understandable that terms and concepts were still being negotiated. The European word Dreaming was used to describe the origin and relationship that exists between humans, the land, and everything else that exists on Earth. It is not a religious concept in the European sense of the word. There are different Indigenous words that are used by different groups across Australia to express this relationship, and because of the difficulties in interpretation and belief, the notion was never subject to detailed discussion in the program. This may have been a mistake in not taking up such an important idea, but it shows how the sensitivity involved and the necessity of showing utmost respect to all Indigenous viewpoints impacted directly on how the program was implemented.

In some respects, similarities can be seen from the idea of evolution that emerged at the same time as the anthropologists were considering how to best describe Indigenous origins. Arising from Mother Earth and returning to Mother Earth can be envisaged as connecting with one aspect of evolution, as well as showing the linkages that exist between all structures of Earth. Being able to discuss such matters with respect and accept that both scientific and cultural views can be recognized and critiqued as humans work together on questions of reconciliation and justice are central to issues of race and discrimination. It would be hoped that in a continuing program, substantial progress in this direction would have been obtained.

Culture as Narrative and Narrative as Culture

As mentioned previously, the course approval submission that went to the Academic Board of the University for approval contained four principles of course design—but the concept of narrative was not included. Our initial writing as background to the course proved to be very accurate in practice although culture, environment, and learning from the land were not strongly emphasized in the early documents (Hooley 1997, 1998). Significantly, it was noted that greater attention than was provided in the approval document would need to be placed on the monitoring and assessment of learning outcomes and the inclusion of innovative teaching practice as the course unfolded. Each of the four principles would need to be interpreted in relation to local circumstances and be changed and adapted accordingly.

The concept of narrative emerged as our experience accumulated. While pursuing a process of inquiry and incorporating the principle of environment, there remained the issue of how the knowledge that was being formulated could be brought together and communicated to others in a form that was consistent with the ways of knowing of the participants and that was acceptable to university procedures. Case study (Yin 2003), case writing (Shulman 1992), and narrative (Elliott 2005) were three of the ideas that seemed acceptable with the following distinctions being made between them.

As we saw it, case studies often involve the collection of extensive observations of a specific situation combined with appropriate contextual information. They are sometimes used to identify variables and processes that can be isolated and studied in greater depth. Case writing was a technique that we had been using at the university in a particular way in our teaching and research involving description of one specific event or example of practice that has occurred and prompts reflection on the part of writer and reader. A case is a very brief, essentially descriptive piece of writing that is naturalistic and informal in style. In contrast, we saw a narrative as a broader concept that brings together the longer and more detailed structure of a case study with emphasis on the specific items of a case. A narrative is therefore often seen to include a sequence of events for the specific purpose of initial interpretation and discussion and can act as the basis for more detailed case and case study writing.

Within this context of our thinking at the time, the concept of narrative that was thought to be appropriate for the program did not need to adhere exactly to that developed elsewhere, but did need to be clearly distinguished from case study and case writing and evolve with its own characteristics to suit the approaches to teaching and learning being used. As a unifying concept, it enabled the entire course to be vigorously conceptualized as an aspect of the continuing narrative, history, and culture of the local Indigenous people—in effect a study of "from the land, to the land," with each subject and topic contributing experience and meaning. Narrative was not something to be grafted artificially onto a formal course for academic reasons, but instead becomes the course, the continuing and authentic dialogue of the community. While this view can be contested, the concept of narrative can also be seen as a process of modern science, where tentative views, hunches, inspirations, and mythologies all combine to help chart the way forward.

As well as recording and communicating, the narrative can generate the speculative guess and wild idea on which humans reflect and ponder and attempt to construct new ideas to reach further than that which is already known. Conceived of in this way, the narrative is inherently respectful and cultural because its evolution depends on the explicit background knowledge, understandings, and interpretations of everyone involved. For Indigenous peoples, such an approach begins with the customs, stories, and wisdom of the community, rather than the traditional, non-Indigenous and elitist literature and texts of the university. This is not to deny the latter, but to ensure that in two-way education and learning, Indigenous knowing and doing remain the starting point. Bringing meaning to the construction and reconstruction of practical knowledge can proceed through community narrative, to which all contribute as equals. There is thus a practical, cyclical development of explanation for community consideration and guidance, rather than a predetermined and constricted theoretical analysis of practice from a non-Indigenous perspective.

These notions of narrative and story did not arise accidentally. They were already embedded in the approaches to teaching and learning that the university staff held regarding integrated knowledge and inquiry processes. They also drew on the importance attached to reading and writing for ordinary people. Reading and writing figure prominently in all areas of the academy and we saw this as continuing in the program. However, we accepted that the extent and format of narrative expression may differ from more traditional modes when Indigenous students are involved. The key issue we thought was not the volume of reading and writing but the manner by which meaning is represented by the various narrative elements concerned. The consensual meaningfulness of a sequence of events can be derived from artifacts, productions, exhibitions, group discussions, as well as a wide variety of writing genre, all of which demonstrate biographical and autobiographical histories and experience. It was agreed that it was a matter for professional judgment as to whether a narrative discussion results in a written case, or a dance and painting, recorded via photographs and video.

We felt that this approach to narrative as we best understood it at the time could be strongly defended. A program that centers on inquiry, environment, and narrative satisfies the dual requirements of intellectual and cultural integrity and recognizes university academic procedures for quality and rigor. All three principles have a strong literature and experiential base and are widely adopted across the social and physical sciences. Taken together, they would establish a framework that involves consideration of appropriate and agreed course outcomes by both the local community and university committees. The educational purpose of the narrative would be an accurate portrayal of experience over time and the respectful hearing of practitioner voices so that meaning can be negotiated and reworked into practical forms that lead to improved teaching, learning, and reflective critical consciousness. We recognized that this was not going to be an easy journey and that much learning would need to occur as we proceeded. We hoped that such a process would be of interest to the entire university if not the entire country as we worked through assessment, accountability, and evaluation mechanisms for a wide variety of observers and critics.

Learning from the Experience

What criteria do we use to judge the success or otherwise of a Bachelor of Education program for beginning teachers who are Indigenous? Particularly one that was established so quickly with good will but limited experience in a small country town? In the first instance, we acknowledge that the local community were extremely proud that their young and mature age people were able to attend an approved university program in the main street of their town and achieve a qualification without needing to leave home and family. Not all Koori graduates took up positions as teachers and in fact, most secured a range of different occupations locally. This enabled a regular income and increased the standing of Indigenous people in the eyes of local townsfolk. It has been reported that those non-Indigenous graduates who gained employment as teachers have contributed markedly to the life and curriculum of their schools to the benefit of Indigenous children. It is always a key question for teacher educators as to how their graduates find their feet in the robust environment of schools in the first years of teaching and to be able to detect an emphasis on Indigenous education is a fine outcome indeed.

From a university point of view, we learnt that a democratic and collegial approach to overall governance was extremely important, although often antithetical to the usual structures of universities. This involved a management committee that had a majority of Indigenous members, staff, and students and that had the attendance of senior university personnel at various meetings. Under these conditions of management, the local community was pleased to participate in decision making regarding policy, but did not get involved in the daily running of the program or the work of lecturers. Mistakes were made by staff in not keeping the channels of communication open at all times on all issues. It was difficult to maintain contact with all families and older school children to encourage ongoing discussion of educational futures and employment pathways, a situation that the provision of mentors on a consistent basis may have alleviated. In a similar vein, there were not always senior university personnel who were familiar with and involved with the program to advocate on its behalf. This was an important consideration when issues regarding the budget were to be discussed. For example, one perception by some community and university members was that the program was essentially white. For community members more directly involved in the program however, there was delight at the entry of non-Indigenous students into the course. This acknowledged the standing of the degree in the eyes of the broader community, as well as providing respectful contact between Indigenous and non-Indigenous students on a daily basis. This was a difficult point for the university to accept entirely, especially in relation to Indigenous funding being allocated to a program that involved non-Indigenous as well as Indigenous students.

Given the complexities involved and the graduation of one Indigenous person per year on average, it is generally agreed by the key players from community and university alike that the program was successful from a social, personal, and educational point of view. Indigenous and non-Indigenous peoples of Australia do not have a long history of living and working closely together, of sharing experience, and of attempting to find solutions to serious problems of mutual concern. Our efforts described in this story at setting up culturally inclusive curriculum at the university level and of working in a two-way inquiry learning fashion throughout is one attempt at reconciling differences of culture and understanding. If we had the opportunity again, we might perhaps introduce the concept of narrative inquiry (Clandinin and Connelly 2000; Hooley and Ryan 2008) more systematically as a way of bringing the ideas and practices of university and community together and of more firmly establishing community learning circles to enable the participation of as many people as possible. While the community was insistent on a quality program as recognized by the white system in Australia, it is of equal significance that we ensure the incorporation of Koori perspective and knowledge within teaching so that Indigenous heritage and history is consolidated. This requires patience and commitment over long periods of time. Formal education programs at all levels that seek to respect and understand Indigenous culture and worldview have a major contribution to make in this regard. Our story is one of life-changing experience, of democratic intent, and of righting past injustice for a more satisfying future. We join hands with our Indigenous friends in looking to the years of struggle ahead.

Notes

1. In Australia, the word Indigenous is taken to mean Aboriginal people who live on the mainland and surrounding islands as well as on the Torres Strait islands between Australia and New Guinea.
2. Approval for publishing the material contained in this chapter has been obtained from the Indigenous community concerned.
3. Koori is an Indigenous word meaning the Indigenous people of south eastern Australia.

References

Australia. 2007. *National report to Parliament on indigenous education and training 2005.* Canberra: Department of Education, Employment and Workplace Relations.
Bruner, J. 1996. *The culture of education.* Cambridge, MA: Harvard University Press.
Clandinin, D.J. and M.F. Connelly. 2000. *Narrative inquiry: Experience and story in qualitative research.* San Francisco: Jossey-Bass Publishers.
Dewey, J. 1963. *Experience and education.* New York: Collier.
Eagleton, T. 2000. *The idea of culture.* Oxford: Blackwell.
Elliott, J. 2005. *Using narrative in social research: Qualitative and quantitative approaches.* London: SAGE Publications.
Harris, S. 1990. *Two-way Aboriginal schooling: Education and cultural survival.* Canberra: Aboriginal Studies Press.
Hooley, N. 1997. *Koori education for emancipatory intent: A collection of discussion papers regarding Koori learning and curriculum.* Melbourne: Victoria University.
———. 1998. *Learning environmentally: A collection of writings that connect Indigenous and non-Indigenous culture, knowledge and learning.* Melbourne: Victoria University.
———. 2002. *Two-way inquiry learning: Exploring the interface between Indigenous and non-Indigenous knowing.* Melbourne: Victoria University.
Hooley, N., and M. Ryan. 2008. *Community knowledge in formation: Narrative learning for indigenous children.* Paper presented at the Annual Conference of the American Educational Research Association, New York, April.
Hughes, P., A.J. More, and M. Williams. 2004. *Aboriginal ways of knowing.* Adelaide: Paul Hughes.
Hume, L. 2002. *Ancestral power: The dreaming, consciousness and Aboriginal Australians.* Melbourne: Melbourne University Press.

IWGIA. 2008. *The Indigenous world 2008*. International Working Group on Indigenous Affairs, Copenhagen. Retrieved May, 2008, from http://www.iwgia.org

OECD. 2007. *Program for international student assessment* 2007. Organization for Economic Co-operation and Development. Retrieved May, 2008, from http://www.pisa.oecd.org

Shulman, J.H. 1992. *Case methods in teacher education*. New York and London: Teachers College Columbia University.

Williams, R. 1981. *Culture*. London: Fontana.

Yin, R.K. 2003. *Applications of case study research*. Thousand Oaks, CA: SAGE Publications.

CHAPTER 6

River Journeys: Narrative Accounts of South Australian Preservice Teachers during Professional Experience

Faye McCallum and Brenton Prosser

Introduction

Many teachers leave the profession within five years of graduation, which is an issue of concern for teacher educators, education researchers, and education policy makers. In Western nations, between 25 and 40 percent of teachers stop teaching within five years of commencing (Hunt and Carroll 2003). Other estimates claim that up to 30 percent leave within three years and up to 50 percent within five years (Goddard and Goddard 2006). In Australia there are high levels of teachers leaving within the first eight years (OECD 2005), with estimates of those lost varying between 30 and 40 percent in the first five years (Ewing and Smith 2003; Ramsey 2000).

Internationally, the age profile indicates many are likely to retire in the coming decade (Moon 2007) and the ageing teaching force in Australia (Hugo 2007) has meant that the supply of quality teachers for the future has become a federal priority. In response, Australian state and federal governments have introduced cash rewards (such as reduced tertiary fees, regional or specific discipline teaching incentives, and advanced skills incentives) in an attempt to attract and retain teachers.

Consequently, the OECD (2005) has identified the retention of quality teachers as a main policy concern, due largely to its impact on quality teaching, the loss of teaching expertise, and a lack of return for investment in teacher training. However, retaining teachers within the profession relies on more than such initiatives; retention also relies on teachers viewing teaching as a life long career.

Research into teachers' understandings of their work identifies a number of factors that are associated with increased teacher resilience and retention, one of which is the adoption of an accomplished or positive teacher identity (Alsup 2005; Ewing and Smith 2003; OECD 2005). This emphasis on teacher identity as a means to improve retention does not confine itself to the research literature; it is also an influential part of the rationale behind many tertiary teacher education programs. It is assumed that if a person takes on the identity of "teacher" as their primary identity through teacher education, then they will achieve a level of accomplishment that reflects itself as positive teacher identity. This means that they are more likely to be resilient and retained within the profession. However, if their identification with the identity of teacher is "weak," they will be vulnerable to leaving the teaching profession.

However, such a view does not adequately explain why so many of those who graduate from teacher education programs (with apparently "strong" teacher identities) have left the profession in recent years. In response, we would suggest that one way of considering this anomaly is by recognizing that the above notion of strong teacher identity relies on a simplistic perspective of the nature of identity and teacher identity development.

In this chapter, we argue that there is no single strong teacher identity that all preservice teachers must adopt by the end of professional experience, rather, identity is a multiple and complex phenomena and there are many intersecting identities that need to be considered. With this in mind, we argue that we need new frameworks for understanding the complexity of teacher identity development and then explain a narrative identity conceptual framework that we have found useful in our work as teacher educators. We then explain how we have used this framework to gain insights into the developing teacher identities of one cohort of preservice teachers within our teacher education program, as well as consider in detail the case of one early career teacher, before we conclude with reflections on the implications of these insights for our understanding of teacher preparation, identity development, and retention.

Narrative and Teacher Identity

Within the field of identity theory, there are a number of different approaches that can be used to conceptualize the multiplicity and complexity of identity. However, in our work as researchers and teacher educators (Hattam and Prosser 2008; Prosser 2006a, 2008), we have found the concept of narrative identity to be generative. Our notion of narrative identity builds on the belief that the reality of the social world is that which is cocreated by humans through dialogue and interaction. In short, if people believe and act as though something is real, it will be real in its consequences.

We find examples of such an approach to reality throughout human history in the use of storytelling as a means to order experience, create new knowledge, and cultivate identity (Clandinin 2007; Stone 1988). Today, narratives continue to be central to interpersonal communication, individual growth, and our identity development in social worlds (Gee 2000; Riessman 2008; Sfard and Prusak 2005). In this view, narrative helps us to understand and express experience, as well as organize and construct reality (Freeman 2002; Riessman 2008), or as Bruner (1987) explains it, individuals become the narratives that they tell about their lives. Thus, narrative is not only a tool that assists us to come to understand the world around us, narrative can also be used to help us come to understand how we view ourselves and how we think others view us, namely our identity. As Riessman (2008, p. 7) says

> The push toward narrative comes from contemporary pre-occupations with identity. No longer viewed as given and "natural," individuals must now construct who they are and how they want to be known, just as groups, organizations, and governments do.

We believe that the performative aspects of narrative are important in the reinforcement and sustaining of identity development.

Leggo's (2005) work emerges from a consideration of how teachers begin their careers with an emphasis on mastering curriculum content and behavior management strategies, but as teachers soon master these skills, they seek deeper resources to help sustain themselves in their work:

> One of my great concerns about teaching is that the demands are so relentless that even the most dedicated teachers often experience burnout, dissatisfaction, ennui, hopelessness and despair. Therefore, I claim that teachers, both beginning and experienced, should learn to know themselves as poets in order to foster living creatively in the pedagogic contexts of classrooms and the larger pedagogic contexts outside classrooms. (Leggo 2005, p. 439)

We believe that Leggo's words apply not just to the traditional genre of poetry, but other narrative genres that emphasize the use of metaphor, the exploration of emotion, the aesthetic use of words, and the importance of public performance (Prosser 2006a; 2008).

Also adding to an understanding of the complexity of teacher identity development is the changing nature of teaching according to broader social, political and economic changes in recent years. By way of example, we look to the work of Sue Lasky (Lasky 2005; van Veen and Lasky 2005) who explains how Northern American teachers who entered the profession in the 1970s (where the rationale of education was primarily about student development) now face a time of managerialism and competition (where the rationale of education is primarily about serving the economy). Her research shows that despite the pressure on teachers to change to fit these new priorities, most of these teachers hold on to teaching identities that rely on the values that were developed during the formative stages of teacher education and early career teaching. Lasky identifies amongst these teachers an ongoing tension between their identities and the "right way" to teach or the "right sort" of teacher (as redefined by recent neoliberal education changes). Her work found that these teachers identified themselves as old, impotent teaching "dinosaurs" because they still held on to these narratives of professional identity, often at their own personal and emotional cost. Yet hold on to them they do:

> For teachers in this study, their core values and notions of professionalism were an anchor in a stormy political and reform climate. (Lasky 2005, p. 913)

It is precisely because teachers hold on to these narratives that they find coherence in changing educational environments. In our research with Australian teachers working in urban fringe communities (Prosser 2006b, 2008), we see a similar commitment to formative narratives. Like Leggo (2005), we found that it was the teachers that demonstrated a clear bond to spiritual, humanist, or class-based metaphors that were able to sustain themselves for long periods in difficult and changing teaching environments. As a result, we are attracted to the role of metaphor within narrative conceptualizations of identity. As Witherell and Noddings (1991, p. 1) highlight,

> Stories and metaphors, whether personal or fictional, provide meaning and belonging in our lives. They attach us to others and to our histories by providing a tapestry rich with threads of time, place, character, and even advice on what we might do with our lives. The story fabric offers us images, myths and metaphors that are morally resonant and contribute both to our knowing and our being known.

Narrative, too, is

> a research process... that consciously focuses on nonunitary subjectivity, self-representation, and multiple storytelling, has the potential to generate greater opportunities for respondents to learn about themselves because they work with the researcher actively to understand how their subjectivities fragment, change, and become transformed. (Bloom 1996, p. 193)

Thus, from the wide range of conceptual tools that are available within approaches to narrative identity, we have chosen to use that of metaphor and interview in our teacher education program. In particular, the metaphor that we have found useful to represent the flow of human existence and the complexity of narrative identity is that of the river. We note, however, that this metaphor has been examined previously, for example, in supporting doctoral students to write (Kamler and Thomson 2006).

For us, the river metaphor is useful because it contains within it the notion that identity is fluid, always producing itself through the combined processes of being and becoming (Riessman 2008). It enables us to consider points along the river at which identity development may run quickly, it may eddy, or it may face rocky challenges. The metaphor of the river also allows us to explore undercurrents and the depth of experiences involved at any given time. By drawing on this metaphor we move beyond the notion of an "accomplished identity" or "potentially changing core identity" (Gee, Allen, and Clinton 2001) to one of "multiple and complex narratives constantly being re-written and re-performed around changing experience" (Sfard and Prusak 2005).

Using the "River Journey" Metaphor and Interview in Teacher Education

Background to the Study

The initiative to be shared was with a cohort of graduate entry preservice teachers within a South Australian primary/middle school teacher education program that used narrative and metaphor to explore identity formation during professional experiences (also called practicum, field experience, or placement, which is conducted in schools). At the time of this research, they were in the final semester of study, which concluded, with a six week professional experience in a school site. In this experience, they would undertake full teaching duties and upon successful completion, would gain Teachers'

Registration and subsequently be eligible for employment as teachers. A final class, held immediately after each professional experience, enabled preservice teachers to engage in meaningful personal and professional critique about their time in schools. It is in this context and as part of such a class that preservice teachers were asked to record their "journeys" using the river metaphor.

Data Collection and Analysis Methods

A blank drawing of a river was provided to preservice teachers, which aimed to "map" their personal-professional experiences from week one to six of the teaching block. They were able to draw, write stories or poems, use captions, and add to the river drawing. An authentic representation of the experience was the only stipulation. When they had completed the task (no set time limit was given as we wanted a free-flowing account), they were asked to swap their river journey with a trusted critical friend and reflect. The hard copies were collected for analysis. Follow-up informal interviews were conducted to exchange dialogue and to explore the preservice teachers' unfolding sense of identity. As researchers and teachers of preservice teachers we were aware that the stories shared were told in particular ways, as Bloom (1996) also states that, "...because respondents are often invested in representing themselves in particular ways, they will tell some stories rather than others and will tell these stories in particular ways" (p. 181).

Together the data enabled a representation of each preservice teacher's professional experience. Each of the river maps were analyzed around the themes of identity, resilience, and emotional labor. In the section below, we present one such river journey, selected because it was indicative of the majority of preservice teachers' experiences of identity formation. Subsequently, data was collected by conducting a narrative interview with the same preservice teacher, twelve months post graduation. A revisiting of the survey, retelling of the river journey, and a reflection, all contributed to her ongoing identity formation. These narratives were analyzed using a model of private, dominant, and cover narratives (Prosser 2006a). This process reminds us of Britzman (1994, p. 54) who argued

> that the problem of identity is a problem of language, and thus a problem of fabrication. This is the work of carving out one's own territory with pre-established borders, of desiring to be different while negotiating institutional mandates for conformity, and of constructing one's teaching voice form the stuff of past, that is, student experience.

The idea of teachers creating identity from language, either dominant narratives (the stories of what students, teachers, and schooling should be) or private narratives (the stories teachers tell themselves about what happens behind the closed door of the classroom, or their past experiences as students), are reconciled by cover narratives (stories to explain the situation to others outside the classroom). As such, this data explored how one preservice teacher struggled with the formation of a personal-professional identity and how the use of the river journey metaphor contributed to her understanding of that journey.

River Journeys: Narrative Accounts of Teacher Identities
Retelling Sam's story

Sam, aged thirty at the time of enrolment in the program, had previously completed a Bachelor of Arts degree majoring in languages and had travelled and worked in Japan teaching English to students aged between four and seventy-five years. When she returned to Australia, she applied for the graduate entry Bachelor of Education in primary/middle education at the University of South Australia. Sam's travel, good grasp of her subject areas, and varied employment experiences indicated her understanding about what it meant to succeed at teaching. To the impartial observer, it could be said that she had a positive or strong sense of identity, as an adult and as a potential teacher, and that she was likely to be successful and retained within the profession.

Her first two professional experiences were in a primary school where she had shown accomplishment and passion in her chosen learning areas. She approached the final professional experience confidently and competently. It was at a large city girls' private college comprised of local, rural, and some international students. Sam would be required to teach languages (French and Japanese) in the middle years (years six–ten). She completed the river journey after the professional experience, and her story, her perspective, is represented in figure 6.1.

Sam used this map to represent her struggle with developing a beginning teacher identity. Although she claimed, she was prepared for the placement she also acknowledged that she was "a little rusty on her languages—had to re-learn them during her school visits—but had no time to brush up due to assessment and behavior management." This mismatch between her perceived identity as a beginning teacher and her actual experience showed that she may have underestimated the demands of teaching that are evident in her written narratives, her verbal accounts, and the observations of others.

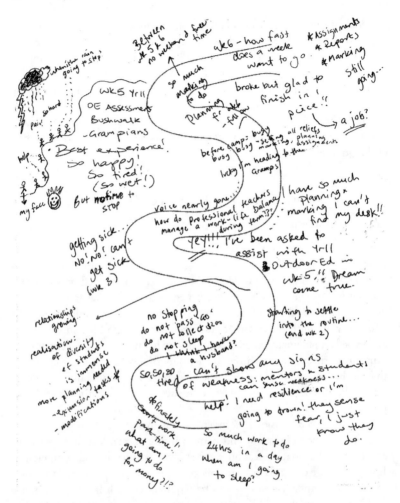

Figure 6.1 Example River Journey
Source: Authors' Data

The river map highlighted many internal and external tensions, and how she negotiated these as she progressed through the experience influenced her own sense of identity and how others perceived her.

The picture in the top left hand corner (see figure 6.1) depicted Sam swimming to the top of the river, which has dark clouds, lightning, and rain as barriers. The catch phrase "when's the rain going to stop?" indicated a questioning of ability to cope. In week one, Sam wrote the following comment:

> So, so, so tired—can't show any signs of weakness: mentors and students can sense weakness... help! I need resilience or I'm going to drown! They sense fear, I just know they do.

This comment indicated early signs of not coping but also an awareness of how she may be perceived by others. This comment also highlighted the importance of self-belief and having a positive sense of identity to enable competent and effective teaching.

In week two, a poem was used to indicate the deepening sense of loss of confidence and an identity at risk of slipping away:

> No stopping
> Do not pass "go"
> Do not collect $200
> Do not sleep
> I think I have a husband?

In this week, there was a critical incident between Sam and her mentor teachers as they disagreed about her ability to teach Japanese and French well. Sam made mistakes during French oral classes because she spoke in Japanese and appeared confused. The mentor teacher informed Sam that she was not coping, because her acquisition of language "was just not good enough." Consequently, her teaching load was changed and all French classes were replaced with Japanese classes. Sam claimed that the mentor teacher expectations were too high for a novice teacher and that planning requirements were far too formal and inflexible. However, she acknowledged the need for the change and wrote "...I fell out of the boat—couldn't teach French and Japanese at the same time!"

The physical demands of teaching became evident as Sam struggled with relational issues, which she described as strong because students responded well to her. Even during week two when Sam's teaching load was altered, she stated:

> They see me as a teacher, just like the mentor. They knew I was struggling with the two languages but were sympathetic and understanding—we would laugh together when I made a mistake and spoke French in the Japanese class.

By the following week, Sam experienced renewed confidence as the change to her teaching load proved to be manageable. She described it as "smooth sailing." However, she continued to struggle with student relationships, as she wanted to be their friend as well as their teacher. She claimed the mentor teachers didn't approve of this relationship, which required her to develop a different nuance to her teacher identity.

In her final week, she completed her professional experience "broke" but not broken, and wondered about future employment. She had worked and reflected on her burgeoning teacher identity in the previous six weeks, but was still conscious that both financially and symbolically, her teacher identity was transient until she became employed as a teacher.

Sam's Reflections on Teacher Identity: Using Interview

An interview was conducted with Sam twelve months after graduation. The interview invited Sam to recount narratives about her early career experiences and developing teacher identity. After two background conversations with Sam, a formal interview of one hour was conducted by telephone because she was teaching in a regional town over five hundred kilometers from the researcher's location. The researcher transcribed the narratives from this interview before thematic analysis was carried out, which focused on key phrases that embodied the narratives in Sam's interview. Using a critical narrative model adapted from Clandinin and Connelly (1996) by Prosser (2006b), they are analyzed according to dominant, private and cover narratives.

- "No mum, this is my home now!" Sam struggled to find employment as a teacher in the first six months after graduation so returned to bar work. However, as the New Year arrived, she was offered a permanent position as a middle school teacher in a rural town. She accepted the position and moved to the town leaving behind a partner, family and friends. A dominant narrative that teachers are involved vitally in the local community (something that was reinforced in her teacher preparation) influenced Sam. Yet, her private narrative was one of dislocation and difficulty. To reconcile these conflicting narratives, she actively tried to build a community identity by staying in the town during school holidays, volunteering at the Marine Centre, socializing with the families of the students she taught, and telling her mother, "No mum, this is my home now!"
- "I'm the only teacher in this school that gives out detention." As Sam established herself as a classroom teacher she experienced some

resistance with students and shared that "they need a lot of discipline at this school, they are good kids, but they test you out." Students tested her so she tried stricter ways to manage them. For example, because she considered written work to be poor she enforced a rule about homework and kept students in detention if they didn't comply, stating that, "I'm the only teacher in this school that gives out detention." She claimed the students didn't like this, or her, but eventually understood and respected her. The dominant narratives are that respected teachers are strict, that good teachers maintain control, that homework is good for you, and that detention or "time-out" will motivate students. Privately she knew she wasn't connecting and may have had doubts because her experience in teacher education told her she should be working on relationships and engagement. But, under pressure, she showed that she was holding on to this "strong," "tough" teacher identity and she would get results!

- "I don't feel that I'm getting any support." Teaching was, at times, difficult for Sam because she taught outside her subject areas, and the small school (sixty-five enrolments) was not well resourced. One day when the researcher rang, she felt despondent after a day where students had behaved badly. She confessed this was because the furniture had been taken out of the classroom where she had arrived to teach. This meant they had nothing to sit on. She recalled, "I can't teach and manage a class when they have nothing to sit on!" Through the conversation, Sam revealed that she had not experienced any induction as a new staff member. She described herself as resilient but was also aware that the school was required to support her. "I am always at them to support me, to give me some time, to mentor me...but it's just so hard up here," she says. The nearest school was eighty kilometers away which made work shadowing impossible, and she couldn't attend any formal induction programs in the city because of the isolation and distance. There were no teachers available to release her which she felt wasn't fair. Sam joined the Teachers' Union who assisted her to clarify her rights and induction options. She felt supported by the leadership team and had initiated a series of formal meetings with the Principal to discuss her issues. She worked hard to prepare lessons for the new subject areas and confessed that she taught herself the content before she taught the class. She declared she also learnt a lot from the lessons that flopped but, maintained, "I don't feel that I'm getting any support." This lived experience is, as Britzman (1994, p. 55) states, "...fraught with ambiguity, ambivalence, contradiction, and creepy detours." The dominant narrative is that education and schools should be well resourced and "new" teachers should be well supported if they are expected to produce quality learning outcomes. Her private narrative is one of disappointment and even frustration because she wasn't getting properly resourced. This manifested itself for student learning outcomes because learning was not supported and therefore, she was not respected.

- "I'm a really reflective teacher" and "I'm not a good teacher yet!" When asked how others perceived her as a teacher, Sam stated, "they see me as enthusiastic, that I care for kids—I love them all—and they probably think I am too understanding or too supportive of the kids. The kids see me as...not good at what I do...they see I'm making mistakes. But the school has a lot of faith in me. They let me try my own things. They trust me." Sam stated "I'm a really reflective teacher" but the cover story is that "I'm not a good teacher yet!" Her private story acknowledged her struggle to perform like a "real" teacher and used a reflective process to help her cope, to be resilient. She persisted, despite having tough days because she believed that a good teacher was a caring one who could make a difference. The dominant story tells us that beginning teachers are allowed to make mistakes as they learn how to teach and this helped to protect Sam's identity as she thought a teacher had acquired subject content, effective student behavior management, sound pedagogy, and positive relationships with others. In this retelling, Sam's accounts showed she struggled with developing her different identities that Britzman (1994, p. 55) describes as "...the slippery relationships among knowledge, experience and the construction of the self" in a context of negotiating "institutional mandates" and developing "teacher voice."

The retelling of Sam's stories highlighted to us their complex nexus and the main themes that appeared in Sam's stories—of her developing teacher identity, her resilience as a new teacher, and of her emotional labor.

Discussion

Unlike most beginning teachers, Sam had the opportunity to tell her story to an academic who cared, and this gave her an outside perspective. If she had told her story to her family, it would have shown a different dimension—perhaps of an accomplished teacher. However, the experiences indicate some of the limitations of a developing teacher identity model. Clearly entering a teacher career with a strong teacher identity has not been enough for Sam, and we would question if this concept provides enough flexibility to understand the complex experiences and identity negotiation of early career teachers like Sam. Further, analysis of her stories shows that for her the struggle to take on a strong teacher identity is creating more hardship than help.

What a narrative perspective of teacher identity can offer is more complex insights into teacher experiences that highlight the implications of these for teacher identity. When Sam says, "I'm not a good teacher yet," in such a difficult environment, we question how long would it be before she stops saying this. When will the paradox between the ideal of a strong teacher identity and her everyday experience become unsustainable and she will leave the profession?

While the narrative identity model helps us understand and describe what might be going on in Sam's situation, it would still appear to be limited in contributing conceptually to helping Sam to respond to her current situation. Perhaps, then, as educators, we should consider that there is a need to move away from the model of a specific teacher identity and descriptions of the complexities of negotiating teacher identity, toward a model that helps teachers think about the teacher they are rather than the ideal teacher to which they are told to aspire to. Fixed notions of identity, whether it is the ideal teacher or the negotiation between powerful teacher identity discourses, may not be flexible enough to think about the complexity, relationships, emotional labor, and constant change in which teacher identity exists.

However, our thinking is challenged by the work of Massumi (2002) and its implications for how we might further reconceptualize teacher identity development. Massumi's focus is around the notion of movement as qualitative transformation and, in particular, a critique of the way that previous cultural theory has emphasized change as the shift between fixed points on a positional grid (e.g., child versus parent, employee versus employer, student versus teacher). He argues that stopping to locate oneself in these binaries can only provide images of change after it has occurred, which leaves the process of transformation relatively undertheorized. If we are to understand human transformation, and in our case identity change, we need to develop sensitivity to the continuity of movement and change, not just focus on identifying artificial beginning and ends.

Massumi explains these general principles in relation to the concept of identity:

> In the everyday intersubjective world, there are of course multiple axes of vision, but they are still strung out along a single line that subordinates them to resemblance and self-sameness. This line is itself nonvisual, it is

a narrative line... You interpret the script, you visualize or form a mental picture of what it means for you to be what you are... For each role there is a privileged other in whose recognition of you, you recognize yourself. You mirror yourself in your supporting actors' eyes, and they in yours. (Massumi, 2002, p. 48)

He argues that this conventional notion of identity remembers only the privileged moments of stasis, but not the movement, which leaves humans aware of progression, but not of transformation. This is "mirror-vision" where you only ever see yourself reflected in one position at one moment, but never effectively in movement as it is "a blur, barely glimpsed" (p. 48). In contrast, Massumi advocates the development of a "movement-vision" that is "unassimiliable to reflective identity" (p. 50). This movement-vision, rather than defining identity in the relationship between "I" and "you," is more self-distancing: "It is to enter a space that opens an outside perspective on the self-other, subject-object axis" (Massumi 2002, p. 51).

Such reflections stretch us as teacher educators because they push us further away from a view of identity that is fixed or is located in a view of the self and other. It also opens up the partiality, complexity, and multiplicity of perspectives within conceptualizations of identity. How we might bring this new view of identity into our work in teacher education is also a challenge that we currently face, but we believe that addressing it will result in rich new insights that will support our work in teacher identity development.

Conclusion

Although our river journey has gained important new insights into identity development and shifted beyond a very singular concept of strong teacher identity, we are still haunted by a sense that we are limited by old ways of thinking about identity. There is still much more to do if we are to understand the multiplicity and complexity of teacher identity development. For instance, we sense that our river journey metaphor is still a linear one; it is a narrative that has a beginning, middle, and end, which by implication climaxes in a strong teacher identity. Further, our use of this metaphor captures a sense of movement, but it still plots a path between fixed and privileged points of reference; it encourages the notion of identity being something that is captured from moments within experience, which belies the idea of constant movement. Third, our use of this metaphor still implies the primacy of one's identity as teacher, rather than a more holistic view of the person and the teacher. In response, we argue that Sam's story points to the importance of new frameworks for understanding the complexity of teacher identity development.

References

Alsup, J. 2005. *Teacher identity discourses: Negotiating personal and professional spaces*. New York: Routledge.
Bloom, L. 1996. Stories of one's own: Non-unitary subjectivity in narrative representation. *Qualitative Inquiry* 2(2), 176–97.
Britzman, D. 1994. Is there a problem with knowing thyself? Towards a poststructuralist view of teacher identity. In T. Shanahan (Ed.). *Teachers thinking, teachers knowing. Reflections on literacy and language teaching* (pp. 53–75). Urbana, IL: National Council of Teachers of English.
Bruner, J. 1987. Life as narrative. *Social Research* 54(1), 11–32.
Clandinin, D.J., and M.F. Connelly. 1996. Teachers' professional knowledge landscapes: Teacher stories. Stories of teachers. School stories. Stories of schools. *Educational Researcher* 25(3), 24–30.
Clandinin, D.J. (Ed.). 2007. *Handbook of narrative inquiry: Mapping a methodology*. Thousand Oaks, CA: SAGE Publications.
Ewing, R.A., and D.L. Smith. 2003. Retaining quality beginning teachers in the profession. *English Teaching: Practice and Critique* 2(1), 15–32.
Freeman, M. 2002. The presence of what is missing: Memory, poetry and the ride home. In R.J. Pellegrini and T.R. Sarbin (Eds.). *Critical incident narratives in the development of men's lives* (pp. 165–176). New York: Haworth Clinical Practice Press.
Gee, J.P. 2000. Identity as an analytic lens for research in education. *Review of Research in Education* 25, 19–125.
Gee, J.P., A.R. Allen, and K. Clinton. 2001. Language, class, and identity: Teenagers fashioning themselves. *Linguistics & Education* 12(2), 175–94.
Goddard, R., and M. Goddard. 2006. Beginning teacher burnout in Queensland schools: Associations with serious intentions to leave. *Australian Educational Researcher* 33(2), 61–75.
Hattam, R., and B. Prosser. 2008. Unsettling deficit views of students and their communities. *Australian Educational Researcher* 35(2), 89–106.
Hugo, G. 2007. *Attracting, retaining and empowering quality teachers: A demographic perspective*. Public lecture, University of Adelaide, South Australia, Australia, April 10, 2007.
Hunt, J., and T. Carroll. 2003. *No dream denied: A pledge to America's children*. Washington: National Commission on Teaching and America's future.
Kamler, B. and P. Thomson. 2006. *Helping doctoral students write: Pedagogies for supervision*. London: Routledge.
Lasky, S. 2005. A sociocultural approach to understanding teacher identity, agency and professional vulnerability in a context of secondary school reform. *Teaching and Teacher Education* 21(8), 899–916.

Leggo, C. 2005. The heart of pedagogy: On poetic knowing and living. *Teachers and Teaching: Theory and Practice* 11(5), 439–55.
Massumi, B. 2002. *Parables for the virtual: Movement, affect, sensation.* London: Duke University Press.
Moon, B. 2007. *Research analysis: Attracting, developing and retaining effective teachers: A global overview of current policies and practices.* United Nations Educational, Scientific and Cultural Organization (pp. 1–45).
OECD. 2005. *Teaching matters: Attracting, developing and retaining effective teachers.*
Prosser, B. 2006a. *Seeing red: Critical narrative in ADHD research.* Flaxton: PostPressed.
———. 2006b. Conclusion. In B. Prosser (Ed.). *Seeing red: Critical narrative in ADHD research* (pp. 283–90). Flaxton: PostPressed.
———. 2008. Critical pedagogy and the mythopoetic: A case study from Adelaide's northern urban fringe. In T. Leonard and P. Willis (Eds.). *Pedagogies of the imagination: Mythopoetic curriculum in educational practice* (pp. 203–22). Dordrecht: Springer Press.
Ramsey, G. 2000. *Quality matters: Revitalising teaching, critical times, critical choices.* Sydney: NSW Dept. Education and Training.
Riessman, C.K. 2008. *Narrative methods for the human sciences.* Thousand Oaks: SAGE Publications.
Sfard, A., and A. Prusak. 2005. Telling identities: In search of an analytic tool for investigating learning as a culturally shaped activity. *Educational Researcher* 34(4), 14–22.
Stone, R. 1988. The reason for stories: Toward a moral fiction. *Harper's Magazine* 276(June), 71–6.
Van Veen, K., and S. Lasky. 2005. Editorial—Emotions as a lens to explore teacher identity and change: Different theoretical approaches. *Teaching and Teacher Education* 21(8), 895–98.
Witherell, C., and N. Noddings. 1991. *Stories, lives, tell: Narrative and dialogue in education.* New York: Teachers' College Press.

CHAPTER 7

Exploring Ways of Promoting an Equality Discourse Using Non-Text/Creative Approaches for Learning in the Everyday Lives of Adult Literacy Learners

Rob Mark

Introduction

This chapter looks at the relationship between literacy, equality, and creativity and their relevance to literacy practice. Drawing on the experience of an action research project on literacy and equality, it examines how these concepts can be linked together to enable tutors and learners to understand equality issues affecting their lives. The medium used to assist this learning is non-text/creative methodologies. Findings indicate that tutors and learners were able to make use of a range of non-text methodologies to improve their understanding of equality and the issues that arise from the inequalities affecting their lives. It also enabled students to develop their own knowledge and skills and their literacies in different contexts.

Connecting Literacy, Equality, and Creativity

If one examines the concept of literacy and what it means to be literate, many different understandings can be found. Popular usage of the term extends from the simple notion of "the ability to read and write" to a host of other ideas including the possession of complex multiliteracy skills such as computer, technical, information, media, visual, cultural, financial, economic, emotional, and environmental skills. A glance at the literature shows that there is no single universally effective or culturally appropriate way of teaching or defining literacy. Rather, definitions of literacy can be seen as a function of social, cultural, and economic conditions. In addition, different discourses may be dominant at different times and in different places.

Throughout the industrialized world, the problem of illiteracy has advanced to the top of the policy agenda, largely as a result of the International Adult Literacy Survey (IALS) (OECD 1997). There has been a radical rethink of the need to confront the issue of illiteracy in national policies, which now recognize the importance of improving literacy for citizens who wish to actively participate in modern, industrial, democratic societies. However, while there is almost complete acceptance that literacy has a profound impact on life chances around the world, there is somewhat less agreement on how adult literacy learning should develop.

Some writers have emphasized the need to move toward an understanding of literacy that encourages critical thinking about the conditions adults find themselves in. For example, Freire (2000), in *Pedagogy of the oppressed*, emphasized the need for "conscientization" of adult learners and more recently, new paradigm shifts have emphasized the need for local everyday life experience to be included in our understanding of literacy needs (Crowther, Hamilton, and Tett 2001). Despite attempts of theorists and practitioners to locate literacy within broader sociocultural contexts, the functional view of literacy as a skill to be mastered still appears to have currency within policy making. Within recent literacy policy documents, there is little evidence of literacy being considered as a critical practice. In many countries, policies refer to the sociocultural relationships that frame literacy, couched in terms of family, community, citizenship, and democracy, but there are few references to the need to examine issues of equality, power relations, and identities.

Models of Literacy

Street (1984) identified two models that can assist with understanding literacy, which he referred to as the "autonomous" and "ideological" model. Each of these models has developed discourses that generate very different ways of thinking about literacy. The autonomous model postulates that literacy is a set of normative, unproblematic technical skills that are neutral and detached from the social context in which they are used. The "correct" skills are defined or fixed (by a powerful group) and learning becomes focused on a mechanical reproduction of correct skills learned in the classroom and which it is assumed may be easily transferred to real life situations. The other alternative ideological model, sometimes called the "social practices model," recognizes the sociocultural, diverse nature of literacy. With this model, power to determine content and curriculum lies primarily with the learner and the social and communicative practices with which individuals engage in their life-worlds rather than an educational organization.

The development of this model to include a "critical approach" adds a further dimension to an understanding of literacy by linking it to social and political issues in society. Shor (1999, p. 15) notes:

> Critical literacy...points to providing students not merely with functional skills, but with the conceptual tools necessary to critique and engage society along with its inequalities and injustices.

Equality Perspectives and Lifelong Learning Policies

Lifelong learning policies in Ireland, both North and South, have emphasized the importance of literacy and basic skills as part of lifelong learning strategies, but with somewhat different emphasis. In Northern Ireland, the lifelong learning strategy emphasized:

> the development of basic and key skills in the context of skills, knowledge and understanding essential for employability and fulfillment." (DEL 1999, p. 1)

In contrast, the white paper on Adult Education in the Republic of Ireland (DES 2000, p. 26) emphasized the need for social cohesion and equity as well as the skills requirement of a rapidly changing workforce in the emergence of an inclusive civil society. The policy agenda is therefore significantly different between the two political jurisdictions in Ireland—Northern Ireland and the Irish Republic—with a particular focus on meeting the needs of the economy in the North and a greater emphasis on equality and social cohesion agenda in the South (Lambe et al. 2006, p. 18).

Working within two very different policy and practice frameworks inevitably posed many challenges for those working in the LEIS (Literacy and Equality in Irish Society) project. However, the emergence of a Peace and Reconciliation process in Ireland, not tied to existing funding structures, provided a new opportunity to work with tutors and learners on both sides of the Irish Border.

A key objective of the LEIS project,[1] was to explore the links between adult literacy and equality issues and to examine how non-text creative learning methodologies might be used to enhance learners understanding of equality issues identified that have affected their lives. The project adopted a social practices model of literacy development that acknowledged the social, emotional, and linguistic contexts that give literacy learning meaning, and that includes a critical approach to literacy. The project held the view that literacy programs should be grounded in the everyday life situations of learners and should embrace issues of equality and social justice. The project brought together a range of people from the field of literacy practice with different types of expertise to promote dialogue about equality as an issue in adult literacy learners' lives.

More than one hundred tutors and learners were involved in the project, which was funded by the European Union (EU) Program for Peace and Reconciliation. Border Action (2006), the funding body for the project, noted that the twin objectives of the EU Special Support Program are to promote the social inclusion of those who are at the margins of social and economic life and to boost economic growth and advance social and economic regeneration. These aims provided a rationale for the project in both jurisdictions. The LEIS project also provided an opportunity to work with tutors and learners in two different political jurisdictions on both sides of the Irish border, sometimes with learners or tutors attending meetings and workshops together from both sides of the border.

Connecting Equality and Creativity to Literacy Practices

Baker, Lynch, and Cantillon (2004, p. 47) note that equality has a complex range of interpretations and, like literacy, is a complex issue to define. In simplistic terms they note that equality is a relationship of some kind or other between two people or more regarding some aspect of their lives. The LEIS project was based on the view that poor literacy skills can be viewed as a manifestation or symptom of inequality and it acknowledged the complexity of the task of helping tutors and learners understand the concept of equality.

The project set out to develop clearer links between a theoretical understanding of equality and practical approaches to including equality issues through the development of creative and non-text methodologies. Using an equality framework developed by one of the project partners, the project examined ways in which creative methodologies could create spaces for the exploration of equality issues within adult literary practice (Baker,

Lynch, and Cantillon 2004). The methodologies were also intended to empower tutors and learners to engage with equality issues relevant to their lives, in particular those arising from the experience of conflict in Ireland.

The theoretical model described by Baker, Lynch, and Cantillon (2004, p. 34) is underpinned by the belief that there are clear patterns in the structure and level of inequality experienced by individuals and groups. The LEIS project focused on four interrelated dimensions of this equality framework as follows—respect and recognition; love, care, and solidarity; access to resources; and power relations. These dimensions provided an opportunity to look at the economic, political, and cultural dimensions of inequality as well as at the affective or emotional realm. The theoretical framework and its connections with the methodological approaches are discussed in greater detail in the projects Resource Guide (Lambe et al. 2006).

Using Non-Text/Creative Methodologies to Explore Inequalities

Non-text/creative methodologies can enable learners to develop an understanding of equality through involvement in a participatory process involving critical thinking and problem solving. Fegan (2003, p. 2) notes that these methodologies can provide a sense of identity and purpose, which can be used to promote greater equality, social justice, and mutual understanding. He also notes they can transform individuals, neighborhoods, communities, and regions.

Greene (1988, p. 125) claims the passivity and disinterest prevalent in classrooms, particularly in the areas of reading and writing, is a result of a failure to educate for freedom. Instead, she argues that we should focus on the range of human intelligences, the multiple languages and symbol systems available for ordering experience and making sense of the lived world. Her theory provides a pluralistic view of intelligences and a holistic picture of how humans learn and can be taught, thus providing further justification for the development of non-text approaches to adult literacy education. Tisdell (2003) also emphasizes the need to take a more holistic view of education, arguing for culturally relevant approaches to adult education, which outline the value of power of "symbol-making and symbol-manifesting activities" and the importance of these cultural experiences through creative activity. In a similar way, Mary Norton (2005) suggests the use of music and visual arts in adult literacy education to provide an alternative way to engage learners.

As the LEIS project unfolded, the need to explicitly emphasize the value of the creative process within each person through access to multiple forms of education became clear. Egan (2004, p. 145) highlights:

> Harnessing creativity is about building on the positive aspects of what is there. It's about drawing on undiscovered skills. It's like a search for gold that, once unearthed, leads to the most explosive release of creativity and excitement.

The creative methodologies used in the LEIS project were used as "codes" to explore equality issues. Adulthood (and consequently adult education) is perceived as a more serious and profound activity than early learning, and as a consequence many adults have temporarily lost much of their capacity to play. This was apparent in some of the responses from tutors and learners to play aspects of the methodologies. The tutors who piloted the methods stressed the value of preparing their students before engaging them in methods that were outside the norm, and to give choice for participation.

The following are some examples of how the methodologies were used in the LEIS Project.

Image Theater

The use of image theater in literacy practice is based around the work of the Brazilian Theatre Director Augusto Boal (1993) who founded the theater of the oppressed, later used in radical popular education movements (Schutzman and Cohen-Cruz 2002). Image theater can help students articulate their own experiences of specific inequalities including situations of conflict whether in the classroom or local community. No ideas need be censored no matter how outrageous or impractical. People who have direct experience of political oppression tend to find it relatively easy to think of images and to make images out of conflict. The spontaneous nature of the improvised image means that they don't have to be perfect. In the LEIS project, the tutors found this method challenging and exciting and considered it to be a useful tool for literacy work especially when exploring issues around literacy, fears, anxieties, shame, and achievements.

Visual Arts

Visual arts methodologies can include a range of activities such as three-dimensional sculpturing and collage. It encourages the learner to play and be creative innovatively without expectations of how things are supposed to be. It enables participants to move into different ways of thinking and doing. They enable ideas and feelings, not always easily accessible to be expressed, enabling issues to be explored and new ideas related in ways that are not easily accessible through dialogue alone. The artifacts produced can be used to engage in discussion, which begins a process of exploring and working with equality issues through concrete and metaphorical ways. It offers an opportunity to deepen the individuals' understanding of one's own and others perceptions and to become aware of similarities and differences that in turn helps build more positive relationships and respect amongst the people involved.

Through the visual arts, participants created concrete artifacts (including sculptures and collages) that represented and communicated experiences and issues. The artifacts serve as tools to reflect upon and describe the learners' experiences in a way not always accessible through words and thinking. One example of how the visual arts contributed to explore equality issues in learners' lives was through the learner creating a sculpture—in this case one of a judge, and using it to discuss feelings of intimidation for someone with low literacy skills in a court of law. These feelings were exacerbated by a legal language and "costumes" or the dress of solicitors. In the focus group, he was able to use the sculpture to articulate his experience of inequalities.

Storytelling

Storytelling is simply the art of telling stories that have been stored in the storyteller's mind. It can include folk tales, myths, fables, personal and community oral histories, and the like. Hardy (1974, p. 13) noted the qualities that fictional narratives play as a major role in our sleeping and waking activities:

> We dream in narratives, daydream in narrative, remember, anticipate, hope, despair, believe, doubt, plan, revise, criticize, construct, gossip, learn, hate and love by narrative. In order to live, we make up stories about others and ourselves about the personal as well as the social past and future. In the same way every person is a storyteller and once they realize this, their confidence and skills in storytelling can improve.

Bruner (1986) recognized that narrative plays an important role in the individual, developing "meta-cognition" or the ability to think about thinking. It is something that can be developed, nurtured, encouraged, and actively learned. It arises out of the sociocultural context in which each individual exists, and this sociocultural context is understood and expressed through stories. It can raise self-confidence and self-esteem and creates a knowledge and awareness of narratives in life, the world, and fiction. Stories can maintain and develop literacy skills through using them meaningfully in learners' lives. Both self-confidence and self-esteem are perhaps building blocks that contribute to understanding the meaning of equality in the lives of learners. It can also contribute to conflict resolution by allowing individuals to have their stories heard. Through listening to the stories of others, they can learn that their own stories are simply one of many and it can enable them to act out alternative endings and come up with practical ways of finding solutions to problems. It can also help develop speaking and listening skills, vital components of any reconciliation process. A comment by one of the tutors indicates the importance and value of storytelling in promoting literacy skills and conflict resolution:

> I began storytelling sessions with my learners by telling my own story. My learners were interested in my story and began to contribute their own. I noticed they became more involved with an increase in concentration and willingness to open up.

Engaging Tutors and Learners through a Participatory Approach

To find out how equality issues might be better understood in the lives of learners, the LEIS project piloted five non-text/creative methods with groups of literacy learners and tutors: The non-text/creative methods were image theater, storytelling, visual arts, drama, and music (the use of a gamelan, which is a musical instrument from South East Asia and which develops skills through equality relationships). A participatory approach, where tutors and learners engage as equal partners was used. Through stakeholder dialogue, the project discussed equality issues seen as important to both tutors and learners.

In the initial phase, both focus and pilot groups consisting of adult literacy learners and tutors from various community organizations, explored issues of equality and inequality in learners and tutors' lives. The focus groups included one hundred tutors and learners and looked at what motivated adults to learn about inequalities, what kind of issues they want to know about, and what would be the best ways of involving adults in this kind of learning. Non-text creative methodologies (the use of collage, image theater, storytelling, and popular theater) were also piloted and provided information on how effective tutors and learners considered these methods to be. These learning methodologies were piloted alongside the equality framework (Baker, Lynch, and Cantillon 2004).

A series of short courses for literacy tutors were also organized around the themes of the project. The courses brought tutors and creative learning methodologists together to work in dialogue with each other, reflecting on various dimensions of equality through a range of activities and examining ways in which learners could be engaged in equality issues. This included a discussion about the links between equality, creativity, and literacy using the equality framework.

For tutors it was important to have an understanding of how inequalities adversely impacted individual lives and to know how to use models and tools to explore equality issues with learners. One hundred and twenty-five people attended the courses, which were organized in seven different locations across Northern Ireland and the Irish Republic. The program included seven continuing professional development courses, each lasting ten hours, and further five courses, where training was part of an initial and ongoing professional development course for adult literacy tutors and managers. Some of the courses included community activists and literacy volunteer tutors lacking formal education and training. Most of the courses were offered as accredited courses and approximately 107 individuals were awarded accreditation. The focus groups and seminars emphasized the need for support materials and resources for tutors and learners. A resource guide that included a rationale and project aims and discussion of the theories and methodologies employed with practical examples of how to use them was prepared alongside the research process (Lambe et al. 2006).

Learning from an Intracultural Research Process

The responses from learners showed that many had learned new skills in communicating and felt more confident talking about the issues affecting their lives. Comments showed that adults with low levels of reading or writing literacy were able to actively participate in learning, thus contributing to the broader goals of social inclusion and citizenship in lifelong learning.

Tutors who used the creative learning methodologies in their practices spoke about the fullness and meaning evident in the level of engagement of learners. As well as encouraging learners to think about issues of inequality—for example, access to jobs, race, religion, and gender—participants also spoke about the methods as being inclusive, encouraging imagination, improving self-esteem, creating a bond between groups, and leading to improved listening skills. Through the use of non-text methods, tutors began to see how they might open up spaces for learners to question previously held assumptions on a range of equality issues affecting their lives as a result of low literacy skills.

One of the tutors commented on how attending training using creative non-text methodologies had helped her develop skills and knowledge in the use of these methods and had also enabled her to see how the methodologies might be used to help students understand the causes and consequences of conflict. Storytelling had enabled her students to talk about real life experiences of unemployment, alienation, and isolation experienced by individuals and communities. It had enabled her learners to talk about the things that united as well as divided them, besides allowing them to develop an understanding and empathy for other's points of view:

> I use story telling with a group of young men. Usually if I ask them what they would like to do they say "we don't know." This time when I asked them to tell stories about their lives, it generated a good discussion about what they were interested in developing and that made a good contribution to equality.

Another tutor who taught on a literacy program for young men took along a sculpture she had made to discuss it with this group. She found that it was a good stimulus to discussion as she was sharing her own feelings and thoughts with the group in a very open way. She felt having a concrete object made it easier to raise more complex issues about equality that were generally hard to do in other types of discussion. It was a good stimulus to get the group thinking, rather than asking the group to write down their thoughts, which she felt would not have worked as it would have limited their thinking to what they could write. Commenting on the use of sculpture to explore inequalities, this tutor stated:

> In the course I met with people from lots of different areas. I felt worried about making a sculpture about peace because I'm not artistic and I didn't want to expose myself in front of strangers. Anyway we worked in groups and it was great because doing it together lead to lots of discussion. We found that what we made together was much more interesting than what we could have made on our own. When everybody talked about what their sculpture represented, you got right to the heart of things because it was a safe space and we were all able to speak honestly.

Another tutor working in a rural college used a collage to encourage students to represent their views about inequalities in their lives. This enabled students to think quite deeply about issues that had affected them without being inhibited by the need to write down their thoughts. She reported that students worked well together and talked about their individual experiences using the collages as a media for the discussion. Issues discussed included a previous lack of education opportunities, feelings of powerlessness in creating change, and a lack of understanding arising from the religious and political division in society.

The tutor noted that:

> Actually participating in making the collage increases the students understanding of equality and enables people to talk about themselves honestly without having to put pen to paper.

Many tutors spoke of the enhanced understanding of the causes of conflict that they were able to link to their own lives as well as the lives of their students, which they were now able to link to their teaching and to helping resolve conflicts. One tutor commented:

> They provided me with simple exercises for conflict resolution which I was able to use with my students. Even my students with learning difficulties easily understood the idea that you need to learn how to cooperate if you are going to solve conflict...

Some tutors spoke of the value of the methods in creating safe spaces for groups from both communities to explore equality issues impacting on their lives which lead to a common understanding of how a lack of literacy skills can created inequalities.

Evidence from the pilots showed that the use of non-text/creative methods with both teachers and students had both educational and social benefits. Tutors reported that their understanding of literacy and how it might be acquired had been challenged. In addition, they demonstrated that the use of these methods could provide a more inclusive way of learning that are not based on text-based forms of knowing, being, and doing.

The methods also enabled tutors to develop new skills and created greater levels of cooperation and understanding between literacy tutors in both jurisdictions of Ireland. One tutor noted:

> I found working with tutors from another part of Ireland made me look at my own practices more openly because what I had taken for granted about accrediting learners was different for them. It was a bit uncomfortable to have the things I see as common sense challenged but it did improve my practices.

Exploring Equality Issues in Adult Literacy Education

Some of the tutors also expressed their criticism of the use of creative methodologies. Some felt the activities might require a high level of preparation or be perceived as "childish" by learners, while others questioned the value of activities having so much fun. For some tutors and learners, education can be perceived as a serious activity where it is not always easy to equate learning as synonymous with a high level of enjoyment. While tutors were very enthusiastic about the use of creative methods for exploring equality issues, they also indicated that ongoing advice and support could be necessary to facilitate the introduction of creative methodologies into literacy practices. They indicated that such advice and support would help build tutors confidence in their abilities to use the methodologies. They also spoke of the need for a clear rationale to validate the learning in the eyes of managers and funding bodies. These comments showed that, while tutors were enthusiastic about the new methodologies, they were also aware of their limitations, many of which were practical, but which could nevertheless be important in determining success.

The seminars provided tutors with an opportunity to generate new ideas around equality issues in the curriculum and in finding new ways of working, using creative methodologies in different situations and environments. The resource guide was also seen as a useful tool for the induction of new tutors, and overall the methodologies used were welcomed as a way of enabling learners to become aware of and talk about equality issues affecting their lives.

When used alongside the equality framework (Baker, Lynch, and Cantillon 2004, p. 34), the new methodologies were found to enable tutors and learners to explore inequalities in new and creative ways. Tutors noted they had learned to recognize the need for respect and recognition of difference, important for each learner. Other parts of the framework enabled tutors and learners to explore the emotional dimension of learning through, for example, the examination of painful issues in people's lives using creative approaches.

Tutors also recognized the importance of access to resources, through limited availability of provision at times and locations to suit learners, but they were very often unable to bring about the necessary changes that would lead to greater equality. Tutors developed a greater understanding of how they could recognize strengths and expertise through dialogue with learners. They found that the new methodologies, with less emphasis on the skills of reading and writing and more on the ability to express views in an open and nonjudgmental way, shifted the balance of power between tutor and learner. Tutors noted a greater understanding of power issues between themselves and students as an important aspect in promoting equality.

The project helped tutors embrace theories of equality and how they could be used to engage literacy learners in a debate about equality in learners' lives. By challenging a "literacy as skills" approach, the framework helped tutors understand and articulate a "nondeficit" perspective that can engage learners in understanding inequalities in their everyday life situations. Through the project's advocacy of creative methodologies, tutors' and learners' understanding of literacy was widened to include visual literacy, oral literacy, and situated learning within creative processes (storytelling, drama, music, and visual arts).

The equality framework was used as a tool to initiate discussion about inequalities. In the ensuing discussions, tutors also raised issues around structural and institutional inequalities that create barriers to using creative approaches, such as the difficulties in working within rigid curricula and the privilege of learning through text-based work. The project also explored potential to examine and discuss "power relationships" through the use of non-text methodologies. A range of issues such as health, housing, welfare, and family was mentioned.

Conclusion

The LEIS project showed how non-text/creative methods can serve as codes to explore issues for initiating reflection and discussion on equality issues, followed with more critical thinking and action. They showed how participants were able to introduce a problem or issue with a purpose of promoting critical thinking and action. The methods, which can promote socially or emotionally related responses, can also lead to deeper understanding of a range of issues affecting learners' lives.

Some tutors saw the equality framework making changes toward a more democratic process, while others saw it as a way of working across national and global networks. Still others saw it simply as a mechanism for talking about equality issues or simply improving the literacy skills of learners.

The LEIS project also demonstrated how working with partners across different sectors can facilitate the integration of new knowledge and ideas that can in turn improve practices. In this case, a teamwork approach enabled a framework for equality to be translated into literacy practices, thus ultimately changing the ways literacy learners think about inequalities in their lives and the lives of others. It also provided new opportunities to involve tutors and learners together in researching their own needs and, in so doing, influencing the development of practices.

Perhaps one of the lasting achievements was building greater insight and understanding of the causes and consequences of inequalities and the possibilities that exist for change. The project's approach to literacy work challenged the widely held view of deficit among learners and instead focused on people's ability to do what they want in their lives. In the words of one tutor, it "opened my eyes and mind to what is possible through using other non-text methods." Through dialogue between tutors and other professionals, the process also promoted greater understanding of equality and literacy practices across the island of Ireland. In addition, the work of the project was based on the premise that literacy is far more than a set of basic skills, rather it is a set of social practices. Adult literacy education is in itself an issue of inequality, since adults with low literacy skills are more likely to be unemployed, living on low incomes and experiencing poor health and early morbidity (Bynner and Parsons 2001; Hammond 2004; Raudenbush and Kasim 2003).

Using a "social practices" account of adult literacy means that instead of being viewed as a decontextualized, mechanical manipulation of letters, words, and figures, literacy can be located in social, emotional, and linguistic contexts. Literacy practices can be seen as integrating routines, skills, and understandings that are organized within specific contexts, and also the feelings and values that people have about these activities.

By focusing on equality and creativity, the LEIS project has shown how theories of equality and non-text creative methodologies can be used to develop new skills and understanding for adult literacy learners. It empowered adult learners to critically examine some of the many issues affecting their lives. Finding ways of addressing these inequalities has no easy answers, but this should not be seen as a reason for denying learners the opportunity to examine and discuss these issues within literacy programs and practices.

The LEIS project has shown how a focus on both equality and creativity can develop new skills and understanding that can empower learners through promoting understanding of inequalities that affect their lives. Speaking about this, Shor (1999, p. 1) argues:

> This kind of literacy...connects the political and the personal, the public and the private, the global and the local, the economic and the pedagogical...

The challenge for the future will be to find ways of aligning literacy policies and practices to the broader goals of equality and social justice.

Note

1. The Literacy and Equality in Irish Society (LEIS) project is an example of a project that used alternative non-text methodologies to help literacy and basic education learners explore and understand how inequalities in society have impacted on their lives. The project focused on inequalities, shifting the emphasis in literacy and basic skills practice away from using printed material to encouraging learners and tutors to explore together the experience of using non-text based methods of learning. The particular focus for inspiring this new type of learning was the postconflict situation in Ireland and the need to understand how an understanding of equality issues could contribute to peace-building. Details about the project can be found on the project Web site (www.leis.ac.uk).

References

Baker, J., K. Lynch, and S. Cantillon. 2004. *Equality from theory to action*. Dublin: Palgrave Macmillan.
Boal, A. 1993. *Theatre of the oppressed*. London: Pluto Press.
Border Action. 2006. Consortium with responsibility for the implementation of the Peace & Reconciliation Projects in Ireland. Retrieved January 19, 2009, from http://borderireland.info/
Bruner, J. 1986. *Actual minds, possible worlds*. Cambridge, MA: Harvard University Press.
Bynner, J., and S. Parsons. 2001. Qualifications, basic skills and accelerating social exclusion. *Journal of Education and Work* 14, 279–91.
Crowther, J., M. Hamilton, and L. Tett. 2001. *Powerful literacies*. Leicester: National Institute for Adult and Continuing Education (NIACE).
Department for Employment and Learning (DEL). 1999. Lifelong learning: A new learning culture for all. Belfast: DEL. Retrieved January 19, 2009, from http://www.delni.gov.uk/acfbb7f.pdf
Department of Education and Science (DES). 2000. Learning for life: White paper on adult education. Dublin: DES. Retrieved February 2, 2009, from http://www.irishtimes.com/newspaper/special/2000/whitepaper/adult_educ.pdf
Egan, J. 2004. Can you paint my soul? Ethics and community arts. In S. Fitzgerald (Ed.). *An outburst of frankness: Community arts in Ireland—A Reader* (pp. 141–47). Dublin: New Island.
Fegan, T. 2003. *Learning and community arts*. Leicester: NIACE.
Freire, P. 2000. *Pedagogy of the oppressed*. New York: Continuum Books.
Greene, M. 1988. *The dialectic of freedom: John Dewey lecture series*. New York: Teachers College Press.
Hammond, C. 2004. Impacts on well-being, mental health and coping. In T. Schuller, J. Preston, C. Hammond, and A. Brassett (Eds.). *Wider benefits of learning* (pp. 37–56). London: Routledge.

Hardy, B. 1974. *Narrative as a primary act of mind.* In M. Meek, A. Warlow, and G. Barton (Eds.). *The cool web: The pattern of children's reading* (pp. 12–23). London: The Bodley Head.

Lambe, T., R. Mark, P. Murphy, and B. Soroke (Eds.). 2006. *Literacy, equality and creativity: Resource guide for adult educators.* Belfast: Queen's University Belfast.

Norton, M. 2005. Welcoming spirit in adult literacy work. *Research and Practice in Adult Literacy* 58. Retrieved January 19, 2009, from www.literacy.lancaster.ac.uk/rapal

Organisation for Economic Cooperation and Development (OECD). 1997. Literacy skills for the knowledge society. OECD statistics, Canada.

———. 2000. Literacy in the information age. Final report of the International Adult Literacy Survey. OECD statistics, Canada. Retrieved January 19, 2009, from http://www.oecd.org/dataoecd/24/21/39437980.pdf

Raudenbush, S., and R. Kasim. 2003. *Adult literacy, social inequality and the information economy: findings from the national adult literacy survey.* Ottawa and Hull: Statistics Canada and Human Resource Development Canada.

Schutzman, M., and J. Cohen-Cruz (Eds.). 2002. *Playing Boal: Theatre, therapy, activism.* London: Routledge.

Shor, I. 1999. What is critical literacy? *Journal for Pedagogy, Pluralism and Practice* 1(4). Retrieved January 19, 2009, from http://www.lesley.edu/journals/jppp/4/shor.html

Street, B. 1984. *Literacy in theory and practice.* Cambridge: Cambridge University Press.

Tisdell, E. 2003. *Exploring spirituality and culture in adult and higher education.* San Francisco: Jossey-Bass.

CHAPTER 8

Team Teaching: Having the Eyes to See the Wind

Dawn Garbett and Rena Heap

In writing this chapter, we have drawn on our experiences as teacher educators in a Faculty of Education at a large New Zealand University. In the Graduate Diploma of Teaching (Primary) science education course, we trialed team teaching as a way to ensure that student teachers were exposed to more than the modeling "exemplary" practice as the means to learn about teaching science in primary schools.

Content knowledge remained a focus for the students and us; however, what was given more prominence through our team teaching trial was the subtext of learning about teaching. This subtext of learning about teaching was addressed by critiquing one another's practice in front of the students to draw their attention to the complex and challenging craft of teaching science, at any level. In this chapter we discuss our experiences and suggest possible criteria for successful future team teaching partnerships.

A metaphor that we have both used in reflecting on our team teaching experience has been that of enabling students to see the invisible.

> That's a great vantage point if he's got the eyes to see the wind!
> (Commentator, America's Cup Yachting Race, 2001)

As landlubbers, the incongruity of this comment intrigued us. A sailor had been sent to the top of a tall mast in order to better see the shifts in wind that might ultimately give Black Magic, the New Zealand boat, an advantage—if he had the eyes to see the wind. We saw a parallel between the wind and the subtext of learning about teaching. What would the wind look like if you could see it? How could anyone see something so transient, fluid, and invisible we wondered?

We wanted to make more explicit to the student teachers those aspects of teacher education that barely ruffle the waters of their consciousness. We wanted them to understand the complexities and challenges of creating a quality teaching and learning experience. We wanted them to notice that what we made look simple was in reality more difficult and sophisticated than it appeared. We wanted to enable them to see the wind.

The science education course in which we selected to trial team teaching had been rated extremely highly by the student teachers in previous years. Modeling exemplary practice and running well-organized, exciting and dynamic sessions has been our standard practice. We were confident of our own knowledge of the science education content. But we were also aware that while the Graduate Diploma in Teaching (Primary) student teachers were keen to learn science content and the superficial aspects of delivering information, we wanted them to see "beneath the surface to the complex thinking and the wealth of experience so crucial in shaping pedagogically meaningful learning experiences" (Loughran and Russell 2007, p. 218).

Team teaching was a strategy that we experimented with in order to unpack the teacher education curriculum for the student teachers *through* our science education teaching. Generally we naively assume that the students will be cognizant of this without us having to make it explicit. Through reflection and analysis of the students' and our own responses we realized that we rarely made many of our teacher education practices explicit to the students, nor in fact did we articulate them very clearly for ourselves.

Team teaching does not allow lecturers to perpetuate a delivery style that posits the students as passive receptacles of knowledge (Letterman and Dugan 2004). This self-study of our teacher education practices has enabled us to reframe preconceptions that center around the simplistic and misleading idea that teacher education is the modeling of exemplary practice and has challenged us both to consider anew our pedagogy.

Anchoring Positions

Traditional teacher education programs work to provide student teachers with content and pedagogical knowledge based on a theory into practice or application of knowledge model. It is assumed that what all teachers have to do is apply this new knowledge in their professional practice and that this will happen in relatively unproblematic ways. This view largely ignores that much of the expert teacher knowledge is embedded in tacit background knowledge developed throughout teaching practice where the competent practitioner acts appropriately in the situation, "on the spot," "in the twinkling of an eye," and "in the heat of the moment" (Roth 1998, p. 372).

This tacit practitioner's knowledge is embodied in specific contexts that deal with the complexities of classroom life, and so to understand teaching, it must be studied in everyday practice. Teacher education programs must provide opportunities for student teachers to formalize this study of everyday practice. In team teaching we were working to create an innovative teaching format that would make this tacit knowledge more explicit to the student teachers and therefore would more adequately prepare our student teachers for their classroom realities.

Most of the few studies of collaborative teaching in the literature have found that it is a win-win situation for the teachers and students alike (Kirkwood-Tucker and Bleicher 2003). However, it is also suggested that collaborative teaching is untenable at university level given the constraints of time, workload, and research demands (for example, Fish 1989).

Team teaching entails two or more educators working together to plan, teach, and evaluate courses. Discussions of team-teaching from lecturers' perspectives can be found in the literature but there were few that related directly to our situation. Indeed there exists a paucity in the teacher education research literature of collaborative team teaching efforts. It is well recognized that working within a university is largely an isolated profession and collaborations among lecturers in university classroom settings are rare (Kirkwood-Tucker and Bleicher 2003).

In aligning our research with existing studies, a variety of models of team teaching were apparent—but no models that mirrored our case study. When considering team-teaching approaches, Kirkwood-Tucker and Bleicher (2003) have presented team-teaching along a continuum comprised of three different models. To the far left of the continuum lies the series model, the most commonly used model, where lecturers present in a series one after the other. Planning, design, and assessment of the course may be shared but there is no team-work, negotiation about, or sharing of the teaching itself. The lecturers are not present in each other's sessions. At the center of the continuum is the alternating model where two lecturers (at least) are present in each session with one colleague actively teaching at a time. The other lecturer(s) may act as participant observers. To the far right of the continuum is the interactive model. Here the lecturers are both present and are both actively teaching as a side-by-side team, equally engaged in the teaching process. This model requires the most social interaction between the lecturers as they are actively teaching together.

Other models parallel this; for example Helms, Avis, and Willis (2005) have the rotational model, the participant-observer model, and the interactive model. Our team teaching style and purpose does not fit neatly into any of these models but rather is a nexus of the alternating model and the interactive model proposed by Kirkwood-Tucker and Bleicher (2003) or of the interactive model and the participant observer model presented by Helms, Avis, and Willis (2005). At times we were both teaching side by side but mostly we alternated the teaching so that while one was teaching the other was acting as a participant observer.

The key point of difference is that the critiquing by the participant observer was articulated to the students in an ongoing dialogue during each of the sessions. We alternated the teaching and the critiquing roles throughout each session. In this way our own study was more closely aligned with Berry and Loughran's (2002) self-study. They taught together in a similar situation to ours and were critical friends to each other, making their teacherly decisions explicit as they taught and reflecting on their practice through e-mail communications.

This case study attempted to provide new and effective pedagogy in which theory and practice were effectively linked (Korthagen, Loughran and Russell 2006). It allowed us to reflect on and improve our practice and to converse with a knowledgeable colleague about our teaching and learning; prompting and provoking dialogue, reflection, questioning, and the willingness to implement more innovative teaching strategies.

Navigating the Waters

Our research, built as it is around understanding our teacher education practices, adds to the growing self-study literature that challenges teacher educators "to describe, articulate and share in meaningful ways their knowledge of teaching and learning about teaching" (Loughran 2006, p. 10).

There is no one way or correct way of doing self-study. As Loughran (2007) writes, "as the research unfolds so the learning through the research influences practice and, because the practitioner is the researcher, practice inevitably changes through this feedback, thus influencing what is being researched" (p. 15). Self-study can both "inform the practice of the teacher educators who conduct it as well as bring a deeper understanding to the larger community of scholars" (Zeichner 1999, p. 11). In order to become professional knowledge, practitioner knowledge must also be made public, shareable, and verifiable.

There are five methodological features identified by LaBoskey (2004) as being important to self-study. They are that it: be improvement-aimed;

have evidence of reframed thinking and transformed practice; is interactive or collaborative; employs multiple, primarily qualitative methods; is self-initiated and self-focused; and "demands that we formalize our work and make it available to our professional community for deliberation, further testing and judgment" (p. 860).

We used multiple methods for gaining different perspectives on the effectiveness of team teaching. We surveyed our students and asked for their responses to team teaching at a mid- and end-point of the course. Their perspectives influenced how we continued with team teaching, but more importantly they lessened the opportunity for us to indulge in self-justification and self-satisfaction with our team teaching practices.

The way that different datasets are generated and changed in response to the research process is typical of many self-studies. We anticipated that journals would suffice as a written record of our actions and reflections on practice but we also took photos and field notes to supplement these. We returned to these photos and notes after the sessions to trigger in-depth discussions. Analysis of the data was carried out within a self-study research framework using a hermeneutic cycle of inquiry (Tobin and McRobbie 1996) where analysis was centered on our post-course reflections of all the data gathered and looked at the strengths and weaknesses of our team teaching efforts.

A year after teaching this science education course together, we revisited our team teaching approach. This led us to think again about our team teaching and prompted us to each write a survival memo, à la Brookfield (1995). Through reading and discussing each other's survival memos we have each reconstructed an understanding of our roles in the team teaching trial. Sharing this new appreciation has been the impetus of further changes to our pedagogy.

Setting Sail

We teach science education courses, in both the Undergraduate (Bachelor of Education) and the Graduate Diploma of Teaching courses, to prepare our student teachers to teach this curriculum subject in primary schools (five to twelve years old). Science is taught as part of the generalist curriculum by the classroom teacher rather than by a specialist in most New Zealand primary schools.

Typically class sizes in our institution for such courses are between twenty-five to thirty-five students. We were scheduled to teach two of four Graduate Diploma classes in the same timetable slot. In an effort to integrate teaching and research effectively in the University culture, we discussed the possibility of researching team teaching by combining these classes and teaching them together. The first six of the fourteen sessions were taught with the two classes combined. There were more than fifty students in this group, a situation that was problematic for students and us, given the practical component and pedagogical emphasis of this course. Consequently, after the first six ninety-minute sessions we split the classes again and team taught both classes in different time slots. As the course progressed we experimented with different strategies. Teaching as a tag-team evolved to become more, or less, scripted depending on our focus for the session. In the two following scenarios we have captured the essence of how an observer may have perceived our team teaching.

> Thursday, 2.50 p.m. Rena scans the room to check that each table has their tray each with the assortment of specimens including a mushroom, piece of liquorice, block of wood, yeast granules, flowering plant, moss, lichen, grass plant and rock. Within the context of "Living things—Plants" we are looking at the importance of children's prior ideas and at different types of diagnostic assessment. Rena has set the students up with a "Plant Continuum" group activity and asks the students to place their assortment of specimens along a continuum from "Least like a plant" to "Most like a plant." Dawn as the participant observer writes notes.
>
> A hub bub of student voices ensues. One group is arguing vociferously about where the yeast granules belong. Overheard from another group:
> "What's liquorice?"
> "Don't know where it's from, but it's added to medicine to stop the foul taste."
> "What?"
>
> Clearly they are engaged, and are exploring their own views—but Rena and Dawn wonder if the students will be able to detach themselves from the student role to see the pedagogical purposes of the task and the myriad of teacherly decisions that they are making. Will the students pause to ask themselves; Where are the lecturers? What are they doing while we are carrying out the tasks? Why are they stopping with that group in particular? When will they know to finish the task? What's the most effective way of pulling the activity together afterward? How will they keep all of us engaged in that follow up?
>
> As Rena sees that the students are close to completing the activity she is also thinking that she has monopolized the teaching so far and decides to

switch roles for Dawn to lead the group feedback of the activity. Dawn's reaction to the change is to think "Yikes it's my turn. I wasn't expecting this. Don't panic!" She starts by asking the students why she is standing right where she is. Both Dawn and Rena take that opportunity to talk with the students about the "teaching spot." Rena explains that with this particular class she often finds herself moving the teaching spot from the front of the class to a little further down toward the back in order to position herself closer to a noisier group of students.

Have they seized the teachable moment and made the point? Have the students seen a glimmer beyond the engaging activities and science content knowledge? Dawn resumes the class discussion of the plant continuum.

Friday morning, 10.00 a.m. Dawn and Rena are planning the Geology session in the office.

"Have you got the gear list?" Dawn asks.

Rena opens her folder of resources and they look at the long list they have used in previous years. Rock samples for identification, classification charts, nails for scratching the rocks, hydrochloric acid, large sheets of paper, felt tip pens, cooking equipment, and ingredients for making edible rocks.

"Right. We should start with a diagnostic. What do you use?"

"I start with a bus stop activity to see what they know generally about rocks. Then I get them to classify the rock types."

"What about reading the book 'What's under the bed?'"

"And the rock cycle?"

"We won't have time to do any cooking at this rate. The students love that."

"What was our science education focus for this?"

They check in the course guide—"Planning!"

"We should be having this conversation in front of the students!"

On that note they ended their deliberations and asked for the technician to prepare all of the equipment they might possibly need.

Tuesday morning, 8.30 a.m. Dawn and Rena are sitting on high stools at the front of the class and welcome the students.

"You start," Dawn says quietly to Rena.

"We were planning this session on Friday and decided that you needed to be part of our conversation. We wanted you to know what we were considering. Where did we start?" Rena asks rhetorically. "We had a look at the resources that we had available and what we had done in the past."

"Yes," Dawn interjects. "I hate doing rocks. They are so boring."

"What?" says Rena, obviously surprised. "I didn't know that. Do you really?"

"Yes, I find rocks really, really boring. How are you going to engage me as a reluctant learner?"

Rena recognizes the challenge. "Right," she says, seizing the teachable moment. "How do we plan for students like Dawn here who thinks that rocks are boring? How can we get her interested?" And then they are away with the class sitting up and paying attention to the complexity facing teachers when planning.

Fair Winds

Over the duration of this research project, we found that team teaching afforded both advantages and disadvantages to ourselves and our students. The major benefit of team teaching has been our increased awareness of the problematic nature of teaching about teaching. Team teaching provided recurrent opportunities for reflection on our individual practice while providing us with opportunities to learn from each other's experiences.

> We gain from each other's experience. This was such a valuable form of PD. It allowed me to challenge and sharpen own pedagogy. I was encouraged to take more risks, try different strategies, be more innovative. (Journal, April 2008)

At a deeper level, we were also challenged to look deeper into our assumptions and beliefs about what it meant to be an exemplary science educator.

> They think that teaching is about delivering information and processes, how to keep neat records, how to write lesson plans, how to transmit information logically, sequentially, but I am trying to get under that level of technical proficiency to something deeper. (Journal, April 2008)

> It made me look at my own practice through a different lens. How can I move away from tips and tricks, fun classes, and modeling good practice to actually making explicit the skills, pedagogy, PCK and teaching

philosophy to enable them to go out and begin their journey as teachers. (Journal, April 2008)

We realized that there was much more to being a teacher educator than we had realized despite our years of experience. It is a complex and challenging task to teach both about science and about teaching concurrently. Adopting new roles shifted our perception of ourselves as exemplary teachers practicing in safe and predictable ways into a space outside of our comfort zone.

> The very act of positioning myself as a team teacher causes me to be more aware of the bigger picture of what I am doing. A wind that keeps nudging me in a different direction. (Journal, April 2008)

By acting as the experienced eyes in each other's classes we were able to make more explicit to the student teachers the teacherly decisions we were making in practice.

> I thought that team teaching might be able to help them to see behind the scenes, to a deeper level of the teaching that was taking place. I hoped that team teaching would be able to strip away the polished performance and allow them to see that teaching is incredibly complex. Not only do I make it look simple, but after all my years of experience and working with student teachers in a tertiary institution, it actually is simple. So team teaching is a way to interrupt that complacency. Team teaching is a way to jolt me out of my normal accomplished role into one of actually having to think about what it is I am doing. (Journal, April 2008)

While we considered these to be the greatest benefits, the students focused more on the practicalities. They commented positively on having two lecturers' experiences and knowledge to draw on as evidenced by their written comments, such as:

> "Team teaching is a double blessing."
> "Better student: teacher ratio."
> "The teaching is more effective with two teachers." (Student evaluation, October 2006)

This is consistent with the literature that suggests that in team teaching each of the lecturers are contributing their unique backgrounds, strengths, and expertise and that this combination of a mix of teaching styles and skills, varied expertise and viewpoints "can produce a synergy in the classroom that is not possible when only one professor is present" (Helms, Alvis, and Willis 2005, p. 30). Furthermore, students are able to develop critical thinking skills as they synthesize multiple perspectives, skills, and styles and relate this to their larger conceptual framework (Dugan and Letterman 2008). However, we had intended that the students see even more than such functional benefits.

Foul Winds

There were also disadvantages, from our own perspective and from the students'. Some of these were simply at a practical level. Pragmatically, we experienced difficulty with timing in class—everything seemed to take longer than we anticipated. It was also time consuming out of class to debrief, plan, organize, and discuss the sessions. For every hour of face-to-face contact, we spent a minimum of two hours preparing, planning, and debriefing. In addition to this time commitment, there were also significant other less tangible costs.

> It is a big time commitment but even bigger than that is the emotional and intellectual commitment to it. You actually have to think about teaching about teaching! (Journal, April 2008)

An analysis of the students' feedback indicates an overall lack of comments about team teaching being an effective strategy to teach about teaching. This could suggest that they did not necessarily engage with the teaching about teaching that we hoped was occurring in the sessions. It could be inferred that the students were relatively impervious to our efforts to make this explicit. As one student commented in a class discussion:

> "We see the point, we're OK with it, can we just move on?" (Student evaluation, September 2006)

We suspect that he and many of his peers wanted us to focus on science content rather than about teaching and teaching science. This could be due in part to the assessment-driven nature of the Graduate Diploma course and typical studentship behaviors that prioritize that which is to be assessed. There was no aspect of the team teaching that was assessed and so from their perspective it took away from the time that we could spend on areas that were in the assessment schedule.

However, there were other comments that indicated to us that some students recognized our efforts to examine our practices more explicitly. For example, we believe that the student who wrote, "I have no criticisms except you are both too good at making things look effortless" (Student evaluation, September 2006) had seen through the façade of us being experts. This comment epitomizes the real tension that we have grappled with in reflecting on our team teaching efforts. We are both accomplished teachers and not only do we make it look easy, it is in fact very easy for us to model exemplary practice. It is the default position to which we would return in times of stress.

An unexpected finding was that despite us having a strong collegial relationship and high self-efficacy, we were both threatened by performing in front of a peer—albeit a trusted and supportive one. Having a trusted peer in the class as a pair of critical eyes made us über conscious of everything we were trying to do in this research project. We both felt as though we were being appraised by the other.

When we moved beyond these feelings of anxiety about being appraised we found an even deeper level of dissatisfaction with what it was that we were actually capable of achieving in the area of making the implicit explicit.

> We perform as teachers for the students in our science education classes and we are expert science teachers. Teaching them about teaching is a lot more mercurial. There seems to be a huge gulf between a series of "how to" to trying to workshop what it is *to be* a teacher. (Journal, April 2008)

> If I can teach with my eyes closed how do I open my eyes to see what it is that I am doing and what it is that I am doing that the student teachers are not seeing? (Journal, April 2008)

We also had our doubts about whether we had the capacity to pick up on that teachable moment when we were in the midst of teaching science.

> Arrghh. I worried that I wouldn't be able to think quickly enough and in fact I couldn't. I didn't see the artful things you did—they were just part of your teaching repertoire and then they were gone and in the flow of things I missed the opportunity to comment. The students wouldn't have noticed, you didn't notice and so it went unnoticed and unheeded in the milieu. (Journal, April 2008)

Where the Winds May Blow

Professionally, team teaching has been a transformative and stimulating experience. We both believe our confidence and skill as teacher educators grew, and continues to grow. However, the cost of the model of team teaching we developed was personally demanding; emotionally, intellectually, in terms of workload, and in terms of time. Furthermore we were unable to quantify any real shift in student understanding of what we were trying to accomplish at this deeper level. We believe that team teaching is only worth the time, effort, and commitment if it is of benefit to the students as well as the lecturers.

To this end, we intend to modify our approach to ensure that the students understand fully the rationale behind team teaching. This will necessitate us focusing on and targeting specific attributes, behaviors, signifiers, and practices within each session. These will be made explicit to the students at the outset of the lesson. We will ensure that the students realize that we will model less than exemplary practice on occasions to emphasize what we want them to see. Our respective roles will be clearly defined to the students with the understanding that these roles require blatant candor and are ones that we have adopted to suit our purposes and their needs.

In weighing up the drawbacks and benefits of team teaching we consider the following aspects to be important factors for other teacher educators who may be considering team teaching. This list is by no means definitive or prescriptive:

- We found our pedagogical similarities helpful in forming a team in the first instance. We had compatible philosophies of education and similar perspectives on teaching and learning. However, we can see that it may be beneficial for the students to have lecturers of contrasting philosophies and practices.
- The teaching roles you each intend to take need to be discussed candidly. In the interests of making the implicit explicit, both need to be active in the classroom; challenging, critiquing, refuting, and illuminating each other's practices.
- Acknowledge and plan for the time this will take, both within classes and outside of classes. It takes more time to unpack the teaching with the students so that they do "get it." It also takes significantly more time to plan, prepare, and reflect collaboratively than independently.
- Be prepared to feel vulnerable. Leave your fragile self-esteem at the door.
- Create a supportive environment between yourselves. Acknowledge and respect that each is playing an indispensable role in the development of a deeper teaching/learning model.

Team teaching has given us the opportunity to continue to learn on the job. We have seen that being a teacher educator requires more than modeling good practice. We know that our students have seen us grapple with how to make teaching about science educational for them, as teachers and learners of science. Using a collaborative approach is a powerful strategy to present to student teachers as preparation for their future profession. They have seen us actively reflecting on and critiquing our own practice in order to better see those elusive winds that drive our learning journeys.

References

Berry, A., and J.J. Loughran. 2002. Developing an understanding of learning to teach in teacher education. In J.J. Loughran and T. Russell (Eds.). *Improving teacher education practices through self-study* (pp. 13–29). London and New York: Routledge Falmer.

Brookfield, S.D. 1995. *Becoming a critically reflective teacher.* San Francisco: Jossey-Bass Inc.

Dugan, K., and M. Letterman. 2008. Student appraisals of collaborative teaching. *College Teaching* 56(1), 11–5.

Fish, S. 1989. Being interdisciplinary is so very hard to do. *Profession* 89(20), 15–22.

Helms, M., J. Alvis, and M. Willis. 2005. Planning and implementing shared teaching: An MBA team-teaching case study. *Journal of Education for Business* 81(1), 29–34.

Kirkwood-Tucker, T.F., and R. Bleicher. 2003. A self-study of two professors team teaching a unifying global issues theme unit as part of their separate elementary social studies and science preservice methods courses. *The International Social Studies Forum* 3(1), 203–17.

Korthagen, F., J. Loughran, and T. Russell. 2006. Developing fundamental principles for teacher education programs and practices. *Teaching and Teacher Education* 22, 1020–41.

LaBoskey, V.K. 2004. The methodology of self-study and its theoretical underpinnings. In J.J. Loughran, M.L. Hamilton, V.K. LaBoskey, and T. Russell (Eds.). *International handbook of self-study of teaching and teacher education practices* (v. 2, pp. 817–69). Dordrecht: Kluwer Academic Publishers.

Letterman, M.R., and K.B. Dugan. 2004. Team teaching a cross-disciplinary honors course: Preparation and development. *College Teaching* 52(2), 76–79.

Loughran, J.J. 2006. *Developing a pedagogy of teacher education: Understanding teaching and learning about teaching.* Oxon: Routledge.

———. 2007. Researching teacher education practices: Responding to the challenges, demands and expectations of self-study. *Journal of Teacher Education* 58(1), 12–20.

Loughran, J.J., and T. Russell. 2007. Beginning to understand teaching as a discipline. *Studying Teacher Education: A Journal of Self-study of Teacher Education Practices* 3(2), 217–27.

Roth, W.M. 1998. Science teaching as knowledgability: A case study of knowing and learning during team teaching. *Science Education* 82, 357–77.

Tobin, K., and C. McRobbie. 1996. Cultural myths as constraints to the enacted science curriculum. *Science Education* 80, 223–41.

Zeichner, K. 1999. The new scholarship in teacher education. *Educational Researcher* 28(9), 4–15.

CHAPTER 9

Enhancing Faculty Commitment, Hope, and Renewal through Developmental Performance Review

Georgia Quartaro and Bob Cox

After extensive internal consultation and external research, George Brown College in Toronto, Canada, embarked on a new Academic Strategy (2004) to guide the college from 2005–2008. The key strategic priority identified in this initiative is "To make excellence in teaching and learning the hallmark of a George Brown College education." As more detailed projects to carry out the recommendations within this new academic strategy were being planned, it was immediately evident that one of the barriers to achieving "excellence in teaching and learning" was the lack of a mechanism for comprehensive feedback and development for faculty who were past the probationary period. The implementation of a new faculty performance review process could create an opportunity for engagement, reflection, and renewal for community college faculty.

George Brown College is a community college located in the heart of Toronto, Canada's largest and most diverse urban centre. Students are drawn primarily from the Greater Toronto Area, which has a population of about five million. The college was established in 1967 and has over 14,000 full-time students enrolled in about 150 programs of study. Most programs are two or three years in length, although the college also offers one-year certificates, four-year applied degrees, and specialized programs. Programs vary widely and include business, performing arts, technology, nursing, early childhood education, culinary arts, and tourism. The college has over a thousand full-time employees, of whom about half are professors. There is also a large part-time teaching staff, most of whom are drawn from the vocational areas students are studying.

One of the primary recommendations in the new Academic Strategy was: "Implement (faculty) performance review that supports *dual professionals*." Research demonstrates repeatedly that postsecondary teaching does not improve in response to feedback from students course evaluations alone (e.g., Marsh 2007), which was the only formal mechanism in place after satisfactory completion of the two-year probation process. Student evaluation feedback tends to be stable for individuals, although it varies widely among them, and it is not related to years of experience. Although college faculty are often resistant to review processes after obtaining tenure, there is no research support for the belief that teaching skills improve spontaneously over time or are an inherent concomitant of expertise in a discipline (Seldin 2003), at this institution or at any other. The desire to place increased emphasis on faculty development and enhancement at George Brown is similar to processes being developed and implemented at other institutions and promoted by the American Association for Higher Education, among other bodies.

Successful focus on faculty development and increased emphasis on teaching depends on a supportive culture within the institution. Mission statements about the importance of students and teaching will have little effect if published research, obtaining grants, or other activities are actually valued more in allocating resources and giving recognition (Seldin 2003). Thus it was important that the implementation of a faculty development process was part of a deliberate culture change at George Brown College. Priorities had been shifting in less formal ways for some time. The college had recovered from a period of financial strain in the mid-1990s and had had significant growth in student enrollment. In many respects, attention was turning back to the college's core mission.

Although a process for faculty performance review had been developed about ten years earlier, it was cumbersome, prescriptive, and excessively time consuming and was never implemented. Other attempts had been made in the interim, but were discontinued, in part because of opposition by the faculty union. However, in a climate of relative fiscal stability and anticipated additional growth, the focus turned again to the centrality of teaching and learning. There was some concern that in failing to provide performance review, the college was devaluing the role of the teacher in creating the learning environment and not adequately supporting the faculty in carrying out this role well. In light of this history, however, it was important that the new process be developmental and center on enhancing and supporting teaching.

There is a process for faculty during the two-year probationary period and this had recently been redesigned to make it clearer and more comprehensive. Once past the probationary period, faculty received feedback from standardized course-based evaluations completed by students at the end of each course, program-based 360° key performance indicators mandated by the provincial government, and informal mechanisms that some faculty created for themselves. There was no process for eliciting feedback from colleagues or from the faculty member's chair.[1] The faculty member's reflection about his or her own teaching was therefore not structured in any formal way, and participation in professional development varied, often focusing exclusively on the faculty member's discipline.

Development of the Faculty Performance Review

The current project was an attempt to create a viable performance review process that would be comprehensive, practical, and would prompt self-reflection. The performance review process was developed by a small working group. They reviewed previous practices at George Brown College and elsewhere, conducted research on models/best practices, and interviewed potential stakeholders. This group, led by the director of Staff Development, established key principles and values to be embedded in the process. The result was a process tied closely to strength-based models of performance enhancement (Cooperrider and Whitney 1987; Seligman 2002), which emphasize identifying and capitalizing on what works well, rather than on identifying and repairing problems. In particular, the process made use of Buckingham's (2007) framework and some related training materials.

The intention was to build a strong community of teachers and learners focused on creating a positive learning environment. The process had to foster engagement and self-reflection. It also had to allow for the differentiation of teaching and learning styles and recognize that teachers bring varied teaching skills and application of industry experience. While there is a strong focus on interactive teaching strategies at the college, it was important that the process not be prescriptive, valuing some strategies to the complete exclusion of others, regardless of the students or subject matter. The process needed to support professors in their role as *dual professionals*,[2] in keeping with the mandate of the college. The performance review process also had to be seen as valuable and meaningful toward achieving this strategic goal of the Academic Strategy. Finally, it had to be manageable and sustainable operationally.

Several beliefs were underpinnings for this project. There was every reason to think that most faculty were effective teachers who had strengths in both delivery and content, so the emphasis was development and inspirational rather than disciplinary or corrective. Some faculty had won major teaching awards and were well known in their disciplines. Student feedback about courses was generally favorable. However, there had been little time and attention focused on the core activity of teaching at an institutional level and there had been no provision for collegial dialogue and reflection about teaching. Of course, some of this happened informally among professors, but it did so without structure, support, or recognition. To achieve excellence in teaching and learning as a strategic objective, there were critical transformational requirements. More effective feedback and acknowledgement was essential for renewal, engagement, and retention.

The performance review model that was developed incorporates four elements (described in more detail below):

1. The development of a teaching portfolio or dossier
2. Peer observation and support within classroom setting
3. Chair observation and support within classroom setting
4. Summary meeting between faculty member and chair

Implementing the Pilot

Twenty-nine professors and twenty chairs volunteered to engage in the pilot. Faculty ranged from one to thirty years teaching experience, with the majority having taught for at least five years. As there was some apprehension across the college about this pilot, assurances were provided to the faculty and to their union that no disciplinary action could ensue from the activities that were part of the pilot. Therefore some care was taken to ensure that the faculty volunteers were believed to be at least average in teaching skill and that there were no significant concerns about their performance. It was critical that this be a positive, strength-affirming process. Any concerns about faculty performance needed to be separate from it, to be dealt with using the college's usual mechanisms for such matters. As is often the case when volunteers are solicited for something new, many of the faculty who volunteered were already known to be skillful professors who usually received very positive feedback from their students on the course feedback surveys and who were highly regarded by their colleagues. A pilot period from September 2006 through December 2007 was established. Throughout the stages of the pilot, participants met to discuss their reflections and observations about teaching and about the process itself. These videotaped discussions were the primary data source for both the individuals' stories and the interplay among the participants' accounts. Feedback from the participants (professors, chairs, and those managing the pilot) informed refinements for the model refinement and identified issues to be resolved so that there would be effective operational and sustainable college-wide implementation starting in September, 2008.

The Performance Review Model

The Teaching Portfolio

The first phase of the portfolio was the creation of a teaching philosophy by each faculty member. They shared these with one another and refined them through discussion. The philosophy was also shared with the chair.[3] Faculty were encouraged to use the Teaching Perspectives Inventory (Pratt and Collins 2001) to help identify it. This forty-five-item online inventory identifies five different stances college faculty might take and yields numerical ratings on each. It provides guidance for reflective practice, encouraging faculty to think about how their beliefs are expressed in their classroom practices and relationships with students. It is somewhat unusual in that it does not promulgate a single model but frames each teacher's unique combination of preferences within a structure that allowed for easy comparison and prompts discussion using common terminology.

The full portfolio was developed over the course of the pilot and was meant to be a portfolio that the faculty member would continue to update and refine after the project ended (Seldin 2003). Faculty were free to design their portfolios as they wished and were encouraged to be creative in developing something that was personally interesting and meaningful. Options included film, blogs, using PowerPoint, podcasts, as well as paper-based portfolios. Online help was available and they were encouraged to consider developing an electronic portfolio using those tools. The portfolio was to consist of conventional material, such as a curriculum vitae and teaching history, lists of publications, presentations and awards, and so on. It could also include other material of the person's choosing that they believed would illustrate the work they were doing, and some included feedback from students, examples of curriculum, and so on.

Peer Observation and Support within Classroom Setting

This portion of the process included a preparatory meeting between two professors, observation of one or more classes by the one taking the role of the observer, and subsequent discussion for debriefing. Professors were free to choose any other pilot participant to be the observer. Faculty were trained before this observation to ensure that they understood how a strength-based model was to be used in observing and giving feedback. This included training in using peer support protocols. Much of this training was delivered by Dr. Idalynn Karre, an expert external to the college. This was done to provide objectivity and further reduce any fears about potential negative consequences for engaging in this pilot, as she has no supervisory authority at the college.

The training focused on setting aside one's own expectations about teaching. Observers were advised to try to see the class they observed through their colleague's point of view, as expressed in the teaching philosophy and subsequent discussions, so that their feedback was framed in terms of what that person was trying to accomplish. While completely bracketing one's own views is probably impossible, this framed the observation through the lens of the teacher being observed and attempted to minimize an internal comparison between observer and colleague that would result in unhelpful criticism, however well intentioned. There was a meeting prior to the observation in which the professor whose class was to be visited described his/her plan for class, related this to the teaching philosophy statement, and identified any areas to which s/he would like the observer to pay particular attention. The faculty pair were free to decide how and if they wanted to explain the presence of the observer to students in the classroom and typically chose to make a brief introduction and explanation. The debriefing session focused on the strengths the professor had demonstrated. The intent was to provide feedback to the professor being observed and also to prompt self-reflection in the observer, who saw a colleague teaching in ways similar to or different from the observer's own practice. The postobservation feedback process was framed using the acronym GROW, for Goal, Reality, Options, and Wrap-up.

Chair Observation and Support within Classroom Setting

A similar process, again with a pre-meeting and a postobservation discussion, was used. Chairs were also trained on strength-based assessment and feedback by the external consultant, using a model developed from Buckingham (2007). Prior to the classroom visit, the chair and professor talked about the professor's strengths, any issue s/he was trying to manage in the classroom, and made a plan for the postobservation discussion.

Summary Meeting between Faculty Member and Chair with a Formal Outline of the Process, Key Findings, and Development Plan

The final phase of the pilot was a meeting between the professor and chair to establish a professional development plan based on the professor's experience in the pilot. This plan would form part of the professor's employee file and be available for review later even if the professor's chair changed. The sustaining mechanisms include yearly updating of the portfolio and a yearly observation by a peer of the professor's choice, using the GROW model. A more complete review, including meetings and the observation by the chair and a renewed professional development plan, will occur every three years.

Results of the Pilot

Faculty generally agreed that stating their teaching philosophy, clarifying it through discussion with others (including the sometimes surprising discovery that their approaches differed), and sharing it with their chairs was one of the most useful parts of this process. Writing the teaching philosophy statement required the faculty member to articulate beliefs about students, about the nature of learning, and about their role in that process. These beliefs usually operate at a level below conscious awareness but have a strong impact on how the teaching is done and what the faculty member expects of students. Seeing oneself as a facilitator of discovery and learning for a class leads to different behaviors and interpretations than seeing oneself as a transmitter of knowledge or a coach for skill development.

Peer Observation

A series of excerpts from the conversation between two faculty after one of them (C) visited M's class, illustrates the GROW principle for feedback about classroom observation.

Goal
> C: I really enjoyed coming into your class, especially because it was on advertising and looking at the illusion that is advertising. And I know that it's a great class to get to teach because it has all the wonderful visual components. You really get the students hooked in with those ideas. They know they are surrounded by advertising and they know they are being sold on certain things. The particular ads that you chose really got their attention and I watched them go through the process of discovering what was behind those ads. You focused them from the beginning. You had those five questions on the whiteboard and those five questions were present throughout the class and it was a great way to get them to think beyond that actual ad when you played the video clips. It was an amazing way to get them to think beyond that two dimensional focus. The questions, I think, were your best

hook in. It was the "playbill" focus for the rest of the class, in the way you talked about the classroom as theater in your philosophy.

Reality
    C: The PowerPoint was meant to give them the analytical background for that session. PowerPoint is great, but I wonder if there was another way to deliver all those tools for analysis. I know there are a lot of steps and the lists are very detailed with technical information. PowerPoint delivers that and I know students like it, even demand to it after the class. But I was wondering if there is another way, because you had such an engaged audience. They were there, even at 8 o'clock in the morning, and they were hungry to tell you what they saw in the ads.

    M: In our preconversation, we talked about theater as a metaphor as it relates to my teaching philosophy. I see the classroom experience as requiring movement and action. There have to be characters and something that needs to be discovered. I'm glad you've raised this. [This led into a discussion of the merits and limitations of PowerPoint for the material he was presenting in this class.]

    M: By its nature, PowerPoint is prescribed, so it stifles the exchange. You don't get that conversation.

Options
    C: I wonder if there is a way around that, to have that information on a single page that could cover the same content in a less prescribed, more interactive way. Now we've seen the ads, now we've gone through the questions on the board, now we're going to analyze the story line and see how that story line affects me or implicates me. They are generating their own critique as they go through. If there's another way to do that analysis, that would be great. It also makes it consistent with the way the earlier part of the class is. It becomes visual and tangible and that feeds their learning.

    [They explored this idea for several minutes, developing an alternative strategy for students to learn this material and discuss its alignment with M's teaching philosophy.]

Wrap-up
    C: I enjoyed the class. It was so much fun and those students were absolutely loving it.

    M: Unfortunately, that time comes when it is "back to the material, back to the notes." I struggle with it all the time and when you identify an occasion to put PowerPoint off to the side, take it. It does become a ritual, as lively as you want to make it, the lights go down, "watch, listen, understand" instead of "engage."

C: Ultimately, they like to see you work with the material and they like to work with it.

M: The irony is, the PowerPoint is not the show. I like them to see the classroom as the stage in which they are active agents, not passive.

Another professor, reflecting on the observation and feedback process, commented:

> If you go in without any preconceptions and you listen to the strengths and you reinforce the strengths, you can lead people to think. In my experience, when you get somebody to talk about their strengths, they will invariably mention about some of their weaknesses. And then you can turn it into a challenge, and say, "Oh, how do you deal with that?" which leads the person to think. And when you lead somebody to think, you lead them to learn something.

Chair Observation and Meetings

One chair who took part commented about an earlier experience in classroom observation, prior to this pilot:

> I come from industry and in industry, it is totally single-focused with single accountability. Here in the college, the accountability and level of autonomy is completely different. I remember the first time I went into a classroom. I was kind of observing and afterwards I commented to the teacher in the most muted possible way. The teacher reacted in a ferocious manner and lectured me. How dare I become involved in curriculum and in commenting about things he had been doing for years?... If I had been shown how to observe and how to give feedback, it would have made a big difference for me.

In contrast, another chair, who brings considerable teaching experience to her administrative role, described her experience doing the classroom observation:

> I was impressed with David's organization. He had all the material for students in files and they picked it up as they came in. They knew the system. The class and teacher were working together as a team. I asked the students about it afterwards and they said they knew what to do, so

> they didn't waste time.... I learned a lot about how David managed the class, managed the material, really engaged the students. He was relaxed but there is always that edge of tension: "Am I covering the material? Do they understand?" that I could see because I have taught. He had his finger on the pulse of the class.

The allocation of time and attention for dialogue about teaching has been an important element in the success of this project. One professor noted that she had never before had an hour to talk about teaching with her chair. They had spoken about operational and curriculum issues, about student problems, about college-wide concerns, and so on, but until this project created the space and necessity to talk about teaching, that conversation had never occurred. Both were delighted to have had that opportunity.

Conclusions

The project has been an enriching experience for faculty, who shared aspects of teaching that are not usually spoken about and are often not conscious. The observations have created dialogue about teaching, and the portfolio development places this work into a broader context for each individual. Themes of renewal, optimism, hope, and commitment emerged. The effect of this focused attention is increased collegiality among those participating in the pilot, as well as awareness of and respect for differences. Individuals commented about the culture shift, making time for reflection and discussion about teaching. One professor, thinking about his hopes for the project said:

> We believe that teaching at George Brown is at least as good as at any other college. Maintaining that quality is very important to me. I know some colleagues are worried about being evaluated and found to be lacking. I am willing to be evaluated and found lacking if it shows me how I can improve.

Experienced teachers say that involvement in this project reconnected them with the joy and satisfaction they initially found in teaching. The opportunity to share ideas about teaching with people who teach in other content areas created a focus on the process itself.

That said, there are improvements that are being made based on the pilot and the subsequent review. The strengths-based philosophy that forms the basis for developmental review needs to be conveyed even more clearly to ensure that the process is positive. Some participants found the portfolio process daunting and time consuming, and more supports will be needed to ensure that this part of the process is successful. Refinements to make the process smoother and to allow more time for it have been added for the first full implementation.

There is optimism, albeit still some concern as this performance development project moves from the pilot stage to full implementation. The volunteers from both faculty and administration were interested, willing to take risks, and reasonably confident about their ability to participate. Others may enter into the process more reluctantly, and some faculty are still apprehensive about potential misuse of the process. There are practical concerns about the chairs' time and about the expertise of some chairs who were never faculty. The full implementation, scheduled over the next three academic years, will bring all full-time faculty into this process. It will create demands for more professional development, more focus on pedagogy, and a deeper understanding of the students' experience. Those who have participated in the pilot will continue to pursue their development, which adds to the momentum across the college. They have consistently described themselves as reenergized and optimistic about teaching. Although none was truly cynical at the start, several said that teaching had become routine and that they were disappointed that work they once loved was no longer deeply engaging. By the end of the pilot, this was different. They were enthusiastic, keen to try new ideas, and particularly attuned to the ways in which they could maintain that enthusiasm through dialogue with colleagues, some of whom they had not even known at the start of the pilot. One of the professors described his own hopes for the next phase of this project, saying:

> We have been the pioneers on this process and there is a point where a buzz starts. It starts a shift in thinking. It starts spilling over to more and more faculty saying, "We want this. We want to do our own personal development as teachers." We can be known as a college where the teaching is far superior to our traditional teaching institution. That would really excite me.

Notes

1. Unlike at many other institutions, chairs at the college are nonteaching administrators who have direct supervisory responsibility for faculty and programs of study.
2. As the majority of programs are vocationally focused, faculty are expected to maintain currency in their both areas of content expertise (e.g., nursing, microelectronics, accounting) and in instructional methods.

3. As this was a pilot process with volunteers, some other administrators, such as deans, and academic directors, participated in the process in the role of the chair to ensure breadth of participation across the college. Most are ordinarily direct supervisors of faculty while a few are not but assumed that role for the purpose of the pilot. For simplicity, the word chair is used throughout this chapter to indicate the administrator in the supervisory role.

References

Buckingham, M. 2007. *Go put your strengths to work: 6 powerful steps to achieve outstanding performance.* New York: Free Press.

Cooperrider, D.L., and D. Whitney. 1987. *Appreciative inquiry: A positive revolution in change.* Reprinted in P. Holman, T. Devane, and S. Cady (Eds.). 2007. The change handbook. 2nd ed. New York: Berrett-Koehler Publishers.

George Brown College. 2004. Toward an academic strategy: 2005–2008. Unpublished report available on request.

Marsh, H.E. 2007. Do university teachers become more effective with experience? A multilevel growth model of students' evaluations of teaching over 13 years. *Journal of Educational Psychology* 99(4), 775–90.

Pratt, D.P., and J.B. Collins. 2001. The teaching perspectives inventory. Retrieved June 27, 2008, from http://www.teachingperspectives.com/html/tpi_frames.htm

Seldin, P. 2003. *The teaching portfolio: A practical guide to improved performance and promotion/tenure decisions.* 3rd ed. New York: Jossey-Bass.

Seligman, M.E.P. 2002. *Authentic happiness: Using the new positive psychology to realize your potential for lasting fulfillment.* New York: Free Press.

PART II

Stories of Hope

CHAPTER 10

"It Gives Me a Kind of Grounding": Two University Educators' Narratives of Hope in Worklife

Denise J. Larsen

> In my future I plan to maybe become a teacher...I might like to be a counselor at a college...I like working with people so these would all make a good career. I am not so worried about my wages. I think that I would like good wages but helping others is really more important. (Larsen 1978, p. 19)
> —Seventh Grade Language Arts Assignment
> "Me: An Autobiography"

University educators rarely speak of the deeply personal connections they hold with their work. Indeed, large institutions can make very difficult settings for the heart-filled quest of being a teacher. Within the university, it is often safer to speak of teaching in terms of efficiency, to speak of the effective application of educational methods that support efficient learning, and to speak of working to meet the needs of students. These aspects of teaching are important. Yet, they also serve as cover stories (Connelly and Clandinin 1995; Crites 1979) of university education, often leaving other deeply meaningful, more personal stories of teaching unspoken. In this study, I sought to learn about what university educators deeply hoped that being an educator would mean. The inquiry was framed around educator hope and stories told by educators when asked to reflect on their deepest hopes in being a teacher. The chapter begins with a discussion about teacher hope followed by a description of the use of narrative inquiry in this study. Next, narrative excerpts from two university counselor educators, Meagan and James,[1] provide windows onto the role of hope in sustaining teacher engagement and well-being. The chapter closes with my reflections on the role of hope in sustaining a teacher's vital personal connection to career.

A Teacher's Hope

For many individuals, the decision to become a teacher is a response to a deep desire to know that one can make some worthwhile difference in the world (Palmer 1998). More than a career choice, teaching is often understood as a vocational calling and is intimately linked with a meaningful sense of purpose in life. Narrative researchers interested in worklife more broadly (Jones 1989; Ochberg 1988) and those specifically interested in teaching life (Larsen 1999, 2004) have explored this rarely discussed relationship between personal hopes and professional worklife experience. They observe that choosing to be an educator is often a response to personal hopes about what life can be and what it will mean to be a teacher.

Hope has been described as the ability to envision a future in which one wishes to participate (Jevne and Zingle 1994). Hope is often manifest as an essential thread in the narratives of educator worklife. It is prospective teachers' hopes for their students and for themselves that provide motivation for choosing to become educators. Once in the classroom, teachers' hopes drive important teaching decisions. Hope exerts a soft yet insistent tug on an educator's worklife narrative, for intimate and deeply held hopes offer each teacher a visionary end point for his or her worklife narrative. Deeply held hopes provide the personal definition by which teachers will understand whether they have spent their worklife energies well.

Teacher hope and teaching narratives are lived out in an educational context. The university workplace is often a difficult setting in which to live out meaningful hopes. Indeed, concern for a vital and effective professoriate has been an ongoing focus of study for decades (Wisniewski 2000). Research repeatedly identifies the academic setting as a difficult working environment characterized by exceedingly high demands, fierce competition, unhealthy relationships, threats to faculty retention, and much personal unhappiness across the academic life cycle (Jago 2002; Johnsrud and Rosser 2002; Karpiak 1997; Lamber et al. 1993; Magnuson, Black, and Lahman 2006; Menges 1996; Olsen 1992; Turner and Boice 1987; van der Bogert 1991). In my own research with six university counselor educators

(Larsen 1999), I observed that the deeply held hopes that drew university teachers to the academy are often lost in the politics and harried details of everyday academic work. With loss of hope, teachers' preferred worklife narratives become ruptured and teachers risk losing the sustaining energy that a personally meaningful motive for teaching offers.

Hope and the loss of hope have profound effects on psychological and physical well-being (Cheavens, Michael, and Snyder 2005; Farran, Herth, and Popovich 1995; Parse 1999). Indeed, research consistently suggests that hope is intimately connected to engagement in life and preferred life outcomes (Benzein, Saveman, and Norberg 2000; Dufault and Martocchio 1985; Herth 1998; Snyder et al. 2002). Though little research has focused specifically on university educators and hope, Magnuson, Black, and Lahman (2006) have noted that loss of hope appears to be based, in part, on cynicism with the peer-review system underlying many academic endeavors. Learning more about hope in university educator worklife appears to hold important information for understanding teachers' worklife engagement.

Narrative Inquiry

Believing that stories of hope may be vital to our understanding of teacher identity and worklife, I selected Clandinin and Connelly's (2000) approach to narrative inquiry. Their metaphor of a three-dimensional narrative inquiry space framed by: 1) interaction (personal and social); 2) continuity (past, present, future); and 3) situation (place) guided the study. In-depth research conversations with two participants, Meagan and James, provided the opportunity to learn about their teaching experiences, workplace context, emotions, embodied memories, and understandings of various choices they made relative to teaching. Meagan and James were selected for this research because they offered the potential for diverse perspectives on hope and university teaching. Meagan was a beginning academic, a woman pre-tenure, while James was an experienced full professor within five years of retirement. Both participants were specialists in counselor education within Faculties of Education at universities in Canada.[2] One participant was a member of a visible minority.[3] I did not know either participant prior to the study.

Initial research interviews were conducted in-person with follow-up interviews taking place via telephone. All interviews were audio-recorded and transcribed verbatim. Throughout the research process, research journals were used to record developing understanding and emerging questions. As interview tapes were reviewed and re-reviewed along with transcripts and journals, careful attention was paid to explicit and implicit

(Larsen, Edey, and LeMay 2007) narratives of hope and teaching as they emerged.

Deeply Held Hope: Stories of Purpose

Research conversations with each participant began with an invitation to tell stories of becoming a university educator. Responding to questions about what each hoped for in an academic teaching career, both Meagan and James offered stories of self and career that had taken shape over many years. Their hopes for teaching at university were tied to memories of early family life, recollections of enjoying school, and recognizing the profound personal meaning that each took from helping others. Stories of hope at work were intimately linked to what each participant had come to love or believe was important in life. To move forward with hope at work meant taking the risk to believe that living one's loves and convictions could be possible in teaching. *Deeply held hopes* provided important motivation and conviction for dedicating one's working life to teaching.

The Hope that Career and Life Will be a Contribution to Others

Both James' and Meagan's narratives of hope in teaching centered on the desire to meaningfully contribute to the lives of others. When reflecting on hoped-for endings to their academic careers, both James and Meagan were quick to respond that they sought a career in which they had benefited others. As both projected forward to a hoped-for ending to their careers, it became clear that an important aspect of hope was the narrative thread it spun in laying a preferred life course. Both Meagan and James knew how they hoped their stories of career might end. James reflected on his hoped-for ending to career saying:

> I would say thank you and I would say it's probably not over and I would trust that even though it might not always seem evident to me that it was useful...That it was a small contribution in the scheme of things but nevertheless a contribution. That some lives I leave better to some degree because of my involvement so. I mean all that I would trust, but mostly I would say thank you...

Likewise, Meagan replied:

> [I would like to say] I opened the door of possibilities for others, for students, and for colleagues too...And, that I was able to set the stage

for people, to provide that platform...or the props that they needed to go on their journey. You know...a sense that they can be all that they can be.

The Hope to Live Many Stories of Self

Both Meagan and James were drawn to university educator life by the diversity that the academic role offered. Both reflected on the importance of being able to fill various roles, including teacher, researcher, clinical supervisor, and counselor. The opportunity to live out different aspects of self at work was an intense hope both shared. Meagan explored this hope with me:

> *Meagan*: I really, really like research I am finding. I really like the interactions with the students and I like the teaching part of it. I couldn't get rid of the teacher in me and I thought, so what is going to offer me the most opportunities to use all of the parts of me? I could see that the...faculty position actually was more rounded. It could offer me more potential or would use more of my potentials and possibilities for myself to develop and to find out who I, you know, who I could be.
>
> *Denise*: Would it be fair to say that, in taking the faculty position, you really had hope that these various parts of yourself and the values that you hold around theory, research, and practice would all be possible?
>
> *Meagan*: Yes...definitely. I hadn't thought about that, I hadn't articulated that before this interview but yes...definitely. Looking back that was probably, that was kind of key...

Is Hope Important?

When approached to participate in interviews about hope and teaching at university both James and Meagan readily agreed, though each voiced misgivings, fearing that they might have little to say about hope at work. They need not have been concerned. Initial research conversations lasted a minimum of two and a half hours and follow-up interviews at least another thirty minutes. During these lengthy research conversations, the participants had much to say about hope and teaching. As I spoke with each participant, both Meagan and James reflected on the vulnerability inherent in speaking the truth of deeply held hopes about worklife. Both felt that stories of self and teaching go largely unspoken unless protected conversational spaces are fostered, spaces in which deeply held hopes about meaning and purpose can be safely shared. For both Meagan and James hope was integral to teaching life and holistically connected to life within and beyond being an academic. In James' words:

> In a way I think what people are asking is your question really of hope. Where is hope? Where is the hope in my life? Where is the hope in our life? Where is the hope in the world? Because if we don't get a hold of that, not that we have like a simple answer to that, but if we don't go there then people do say "what's the point?" Right. People, people give up...So I mean, so [hope] may appear for some like a bit abstract really or elusive...yet it's so central and integral.

Threats to Hope in Academe

Deeply held career hopes pull the teacher forward into worklife engagement. Threats to deeply held career hopes strain the narrative threads that engage teachers in worklife. When these hopes are threatened, it feels as though energy is siphoned into the struggle to hold onto hope at work. Both Meagan and James shared ongoing and serious challenges to hope at work. They struggled with dominating institutional cover stories of success and fought isolation engendered by lack of community, while they sought to live out their values in a context where institutional values often seemed madly askew.

Institutional Cover Stories and Success

Demands on faculty to embrace institutionally defined criteria of success were identified by both James and Meagan as threatening to hope. Both participants repeatedly identified values and goals enforced by their academic institutions that ran contrary to their own hoped-for visions of worklife. For Meagan, a new untenured faculty member, hopes for her own career were further threatened by a lack of certainty about her career future. She sensed that to voice dissent over institutional values could threaten her own career survival.

> It is very difficult actually...I can see where I began quite innocently—thinking, "Oh, this will be very doable." But not realizing the complexity of the position and the fact that this is up to me...What I am finding is that you are constantly multitasking and there is never really a sense of doing the best that you could because it seems like you can't really focus on one thing all that long...My hope had shifted a bit you know,

realizing that I can't always give it the best and that to me is very disappointing... And, I must admit there have been times... that I have gone through since about April until about now, that kind of a bitterness or a loss of hope at times... it was almost into a depression. I have found myself going down you know... The point is, as I said the VALUES. I am working in a system where the values are not the same as my values. And, so what does that do to one's hope?

Attending to the details required for career advancement, contract renewal, tenure, promotion, and annual increment competitions can dominate an academic's activities and, most importantly, can become defining aspects of one's work and one's self. Feeding a tyranny of the urgent, these deadlines direct academic energy toward institutional demands that may provide job security but little professional sustenance, answering the only most basic hope to keep a job. James saw these institutional demands as external rewards that provide little meaningful satisfaction about the contributions one deeply hopes to make during life. Both Meagan and James identified the demands of tenure, promotion, and annual increment competitions as fostering a focus on self-advancement and placing an emphasis on personal and financial benefits that bore little connection to the deeply held hopes that drew each to an academic career.

Meagan believed that early academic life required her to surrender deeper worklife hopes—at least temporarily—in order to survive tenure review. She became acutely aware of the need to be strategic about her choices and her career decisions. To do otherwise could fatally imperil her university teaching career. She experienced a fear-driven push to become what the institutional tenure process defined as success. Meagan struggled with worklife outside the classroom. To use Clandinin and Connelly's words, Meagan saw the university "as a place littered with imposed prescriptions" (1996, p. 25), prescriptions that threatened to eclipse her very reason for choosing a faculty position. She believed that her occupational survival required her to live a cover story (Olson and Craig 2001) based on what she understood the institution expected of her. Nevertheless, her hoped-for narrative of worklife was not easily surrendered. The promise of meaningful relationships and her students drew her back to her own deeply held convictions, her hopes for what her worklife might be. As she said,

> Then when I meet with students and start back in that teaching mode. I can't sustain that [singular driven focus on career advancement] because that is not really me anyways, you know, so I am trying to find that balance between the strategy and being strategic and being who I am as well...

Lack of Community

Like Meagan, James lamented the presence of institutional cover stories dominated by what he saw as individualism and self-aggrandizement. Acutely aware that his hope was sustained by community, the support of others who believe in him, and knowing that he cherished and was cherished by others, James reflected on the ways in which the lack of community in academe threatened his hope. According to James, community is often feigned within the academy but rarely practiced.

> *Denise*: Do your notions of community and that of hope have bearing on working in an academic institution?
>
> *James*: I think in my early life in the academy I worked really hard to create community because I really saw it as the way to go. And, from time to time over the years there have always been a few colleagues with whom you create community. But I've realized over time (and it didn't take a lot of time), that the institution, like many other institutions, mitigates against community. It really does. So you know it's still the single authored refereed...paper that counts more than the collaborative effort.... So community in...the academy is like a counter cultural notion.... But what I learned early on is that real community, community that matters to me, I had to find outside of the academy. And that would bring more hope in my work in the academy.
>
> So it's an interesting time in a way in the academy, but it's a challenging time for a lot of people because it can be very lonely and isolating I think for people. And when you're in situations of loneliness and isolation you don't sense much hope, you don't feel much hopefulness about the future. You might feel hopefulness that your career will be advanced, but for me (chuckle) that's not much of a, that's kind of a limited notion of hope. I was thinking of hope as larger and broader than that.

Sustaining Hoped-For Narratives in Academe

While threats to hope were clear in both James' and Meagan's narratives, the explicit cadence of hope was also audible. Both participants were able to identify the threads that kept hoped-for worklife narratives alive in the face of strong institutional counter-narratives. Both Meagan and James indicated that they had never consciously identified stories of hope in their worklife prior to our research conversations. Yet, to do so seemed to anchor them in what mattered most to them at work.

Mentorship: A Circle of Hope

Stories of mentoring offered rich sources of hope. Meagan drew hope from the life stories of experienced female academics she admired. These women's stories became her stories to live by. The examples her mentors set and the relationships they offered her helped her believe that she, too, could live her deeply held hopes within a university setting.

> I am thinking of...one supervisor...what I saw was her perseverance and just her, you know, stick to it and the fact that she was a young female in her department you know she has been a real role model and an inspiration for me...
>
> [Also, my former supervisors] are able to do the caring part. They can do the kind of connecting with me and with the wider community...It's what, I guess for me is that hope, that possibility, of being all that you can be and they are able to. You could see them on that path...It is something I can aspire to see that...Those are the things that I value.

Where support for Meagan's hoped-for worklife narrative in academe was strengthened by her relationships with mentors, James said that he found hope in mentoring new faculty. Where Meagan took hope in seeing her mentors model important personal values, James sought to model the same to his own younger colleagues.

> Working with colleagues, especially with younger colleagues in counselor education...At my stage now I'm seen more like a mentor. Working with them and modeling life, academic life, to some degree is a little bit satisfying and hope filled, you know cause I think they, they're kind of interested in, "How do you do that, you know how is it?" I mean some people would see me as a bit of a rare academic. Not unique as in special, but unique in terms of degree of satisfaction or congruence between who I am and what I do. Over a long period of time I've been able to sustain this kind of interest, hope, and really enthusiasm for students and for the program and work. So you know, I suppose that's a little bit you know hopeful too.

Mentoring relationships provided an atmosphere for safe spaces and the creation of *knowledge communities* (Olson and Craig 2001), meaningful to both Meagan and James. Mentorship provided a place for telling extended worklife stories with trusted colleagues. Similar to the descriptions of a teacher-educator who participated in Olson and Craig's research, Meagan's and James' mentoring relationships provided space for continuing conversations that fostered personal and professional knowledge development.

Relationships with Students

When discussing their work with students, both James and Meagan described moments when hope shone—times when they knew that they were living their hope to make a difference in the lives of others. As James put it:

> When I see a graduate student come to terms with something or I see a graduate student develop a newfound skill that gives me a lot of hope.
>
> I'm always amazed each year at the different chemistry in a group and, yet, that interaction in the classroom really does sustain me. I mean it is with them, for them, by them that hope is engendered. You know it's such a privilege to work. It's really difficult to share I think with non-academics what this is like. But, when you think of I am doing, what I need to be doing, and I am paid for sharing with my students what I am most passionate about, this boggles my mind really.

Relationship with Self

In addition to teaching relationships, time spent intentionally nurturing a relationship with self was essential to sustaining hope as a university educator. Moments of quiet reflection and opportunities to write became sources of hope. These sources of hope offered periods of transcendence, an internal space beyond the harried everyday existence of deadlines and administrative pressures. These quiet moments offered space to commune with self. Reflective time was essential to hope and provided opportunities to "be" in the moment with no specific task or goal at stake. James commented,

> I have a daily practice of quiet time and meditation and prayer... So that is enormously helpful for me because it is a centering exercise and it does provide a perspective I think that I wouldn't normally have. My worklife is very, there are a lot of demands on my time and a lot of, you know, it's involved work. So in order to have some perspective of that as a contemplative person I need to withdraw from the activity and be quiet... reflect on "OK, you know, where are you?"

Meagan echoed the importance of time for quiet reflection in supporting her sense of hope:

> I have hope when I have been affirmed in my convictions in some ways. So for me this summer it was more the ability to have some time on my own alone here to do that self-reflective piece and to realize that there was a direction...I didn't have a sense of hope, I didn't have the sense that I saw possibilities for myself in the future. [This summer] I was kind of forced to really just sit with myself...I think the hope began to be developed again...and knowing that there will probably be ups and downs in any career path that I can still be useful and productive in that way.

Creating Hopeful Work Spaces

Creating internal space for reflection was only one way of sustaining hope for James and Meagan. Both participants also shared ways in which they intentionally sought to create externally hopeful work spaces as well. They spoke of symbols of hope purposely placed in their offices. These symbols included small gifts from students and poems. Symbols of hope served as visual reminders of the stories that give an educator hope, including stories of meaning-filled experiences with students. In creating hopeful work spaces, attention was given to fashioning inviting spaces meant to support life, vitality, and relationships. Plants, colors, and images all figured in the creation of hopeful working environments. During our conversations, James shared several symbols of hope that he had intentionally placed in his office.

> [This small handmade gift from a student] is on my filing cabinet so it's actually the first thing you see when you come into my office. There are a few other things there but that one is definitely visible and often people will ask about it.

Meagan also commented on the impact that her office building and physical environment has on hope.

> But I find that physically the environment is not set up for connecting. It is set up to be isolated—long halls with rooms, you know. Often you will see doors closed you know, because the sound echoes. It is really hard to get any work done, if you don't have your door closed. Any ways that we can connect adds fuel to hope.

She saw the opportunity to create more alive spaces as an invitation for hope at work—in some ways, an invitation to bring a fuller narrative of self to work. It was about making space for the academic, the communal, as well as the domestic.

> So I am thinking that you know, I want to talk about a little act of resistance that adds to hope that I have tried to instigate...I developed my room [office] so it has plants and a rug and various things. People have always commented on how homey it looks...it is a small act that can change things. Because when you have pleasant surroundings, it reflects who you are in some ways...

Hope Offers Grounding

In the face of the loneliness, confusion, and overwhelming demands of an early academic career, Meagan recognized the importance of hanging onto deeply held hopes. She also saw holding onto meaningful hope as difficult for untenured faculty. With this in mind, she offered one final piece of advice for those contemplating an academic career. She admonished,

> Be clear as to your purpose, what brings you there, how you should focus in being there and use it as a guide.... Maybe use [it] as a model or as a mission statement or something so that you can hold onto because in all the demands and the business and the expectations...I found it easy to lose track of that and so perhaps having that be clearer for you and something to hold onto because you will come back to it...I think it is [connected to hope]. It is almost like a guiding light that you need and when the way gets dark you need to have, to be able to turn to that again.

Closing Reflections on Making Hope Explicit in Teaching Worklife

Qualitative research interviews often provide a welcoming space for rich storytelling and the exploration of experience. In agreeing to explore their teaching worklife through the lens of hope, both Meagan and James identified hope as an important aspect of being teacher and counselor educators. Each also acknowledged that they taught about hope more implicitly than explicitly. In other words, they attempted to create experiences that supported students' and clients' hope but rarely spoke about hope directly. Meagan's and James' reflections on the importance of hope are mirrored in recent research literature that highlights hope as a key factor in mental health, change, and learning (Cheavens, Michael, and Snyder 2005; Irving et al. 2004). In addition, their use of hope, as an implicit tool, to support student learning and facilitate client change appears to be the most common way in which hope is currently addressed by both educators and counselors (Larsen, Edey, and LeMay 2007).

I am often struck by the ways in which the research interview experience itself can lead beyond a mere recounting of past experience and into new discoveries. New discoveries were certainly part of the interview experience for James and Meagan as they reflected upon hope in worklife. Both participants said that the experience of speaking about hope explicitly when reflecting on their worklife had left them with elevated spirits. In a university context that is often lacking in community, I believe that our research conversations became a safe space to speak of workplace hopes and struggles.[4] Craig (1995) calls these safe places "knowledge communities", open conversational contexts that are often created outside of the formal education setting. Craig (2002) further notes that her own experiences as a researcher in educational contexts became safe spaces to create knowledge community relationships.

Understanding hope in academic worklife required that both the highs and lows of being an educator be acknowledged. Meagan and James' interview conversations traversed a worklife landscape that included experiences of both great joy and painful struggle. In addition, the experience of speaking about hope explicitly during our research conversations appeared to put words to experiences that otherwise may have gone unspoken. During our research conversations, naming the presence and absence of hope at work seemed to have the effect of attuning the ear and sharpening the eye for the appearance and role of hope in each participant's worklife story. Follow-up interviews revealed the impact that a heightened awareness of hope had had. For example, James reflected on the ways in which he had begun to notice use of the word hope. As he put it, "It's like owning a red car. Once you have one, you notice them everywhere!"

Further, in the face of work place difficulties, a focus on hope helped Meagan and James firmly anchor themselves in aspects of work that were most personally meaningful. They both expressed surprise at how much they had to say about experiences of hope at work. They had not expected that they would be able to speak at such length about hope. The anchor they found in their own deeply held stories of hope surprised them. Speaking these stories in a safe space revealed a meaningful foundation for what worklife could be. The interviews created space to speak one's story, to hear one's preferred story, to be witnessed by an appreciative other, and to have that personal professional hope story valued. To support valuing deeply held hopes at work, Meagan encouraged that personal hope stories of worklife be written, especially to strengthen new academics' connection to the personal stories that drew them to professorship. She believed personal hope stories of work were an important grounding or anchor when faced with inevitable powerful institutional cover stories.

Finally, talking about hope explicitly in conversations about worklife may help to render the "soft" aspects of teaching, aspects that are life sustaining but seemingly ephemeral, more accessible to language. Deeply held hopes appear to be connected with some of the noblest values an educator possesses, values that each educator would most hope to live out. By making the language of hope explicit in worklife, we may have the power to strengthen this dignified worklife plotline—a plotline that hope pulls forward in educators. As a researcher and a teacher educator myself, I cannot help but reflect on James' and Meagan's stories of hope. I find myself wondering how teaching, educational institutions, and worklives might be experienced if we found safe spaces to explicitly learn about and support our colleagues' deeply held hopes—and our own. It seems that hope may play an integral role in the knowledge communities (Craig 2002) we seek in teacher education.

Notes

1. Pseudonyms have been used throughout this text and identifying information has been altered.
2. This study has received ethics approval from the University of Alberta Education, Extension, and Augustana Research Ethics Board.
3. Visible minority is a term defined by the Canadian Employment Equity Act and used in Canada commonly to describe persons who are non-Caucasian in race. The term is a demographic category used by Statistics Canada in conjunction with Canadian multiculturalism policies.
4. While "safe spaces" as a classroom metaphor is worthy of careful critique (see Boostrom 1998), I believe that the term offers a useful way to characterize much of the open, reflective, critical dialogue offered by participants in this study. In addition, the necessity of "safety" for professional development and growth is common language within the field of counseling psychology (e.g., Emerson 1996).

References

Benzein, E., B.I. Saveman, and A. Norberg. 2000. The meaning of hope in healthy, nonreligious Swedes. *Western Journal of Nursing Research* 22, 303–19.

Boostrom, R. 1998. "Safe spaces": Reflections on an educational metaphor. *Journal of Curriculum Studies* 30, 397–408.

Cheavens, J.S., S.T. Michael, and C.R. Snyder. 2005. The correlates of hope: Psychological and physical benefits. In J. Elliot (Ed.). *Interdisciplinary perspectives on hope* (pp. 119–32). New York: Nova Science Publishers.

Clandinin, D.J., and F.M. Connelly. 1996. Teachers' professional knowledge landscapes: Teacher stories—stories of teachers—school stories—stories of schools. *Educational Researcher* 25, 24–30.

———. 2000. *Narrative inquiry: Experience and story in qualitative research*. San Francisco: Jossey Bass.

Connelly, F.M., and D.J. Clandinin. 1995. Teachers' professional knowledge landscapes: Secret, sacred, and cover stories. In D.J. Clandinin and F.M. Connelly (Eds.). *Teachers' professional knowledge landscapes* (pp. 3–15). New York: Teachers College Press.

Craig, C.J. 1995. Safe places on the professional knowledge landscape: Knowledge communities. In D.J. Clandinin and F.M. Connelly (Eds.). *Teachers' professional knowledge landscapes* (pp. 137–41). New York: Teachers College Press.

———. 2002. A meta-level analysis of the conduit in lives lived and stories told. *Teachers and Teaching: Theory and Practice* 8, 197–221.

Crites, S. 1979. The aesthetics of self-deception. *Soundings*, 126–38.

Dufault, K., and B.C. Martocchio. 1985. Hope: Its spheres and dimensions. *Nursing Clinics of North America* 20, 379–91.

Emerson, S. 1996. Creating a safe place for growth in supervision. *Contemporary Family Therapy* 18, 393–403.

Farran, C.J., K.A. Herth, and J.M. Popovich. 1995. *Hope and hopelessness: Critical clinical constructs*. Thousand Oaks, CA: SAGE Publications.

Herth, K. 1998. Hope as seen through the eyes of homeless children. *Journal of Advanced Nursing* 28, 1053–62.

Irving, L.M., C.R. Snyder, J. Cheavens, L. Gravel, J. Hanke, P. Hilberg, and N. Nelson. 2004. The relationships between hope and outcomes at the pretreatment, beginning, and later phases of psychotherapy. *Journal of Psychotherapy Integration* 14, 419–43.

Jago, B.J. 2002. Chronicling an academic depression. *Journal of Contemporary Ethnography* 31, 729–57.

Jevne, R.F., and H.W. Zingle. 1994. *Striving for health: Living with broken dreams*. Edmonton: University of Alberta Press.

Johnsrud, L.K., and K. Rosser. 2002. Faculty members' morale and their intention to leave: A multilevel explanation. *The Journal of Higher Education* 73, 518–42.

Jones, W.P. 1989. *Theological worlds*. Nashville: Abingdon Press.

Karpiak, I.E. 1997. University professors at mid-life: Being part of...but feeling apart. In D. Dezure (Ed.). *To improve the academy* (v. 16, pp. 21–40). Stillwater, OK: New Forums Press and the Professional and Organizational Development Network in Higher Education.

Lamber, J., T. Ardizzone, T. Dworkin, S. Guskin, D. Olsen, P. Parnell, and D. Thelen. 1993. A "community of scholars?": Conversations among mid-career faculty at a public research university. In D.L. Wright and K.S. Lunde (Eds.). *To improve the academy* (v. 1, pp. 13–26). Stillwater, OK: New Forums Press and the Professional and Organizational Development Network in Higher Education.

Larsen, D.J. 1978. Me: An autobiography. Unpublished manuscript.

———. 1999. *Biographies of six Canadian counsellor educators:* Stories of personal and professional life. Ph.D. diss., University of Alberta.

———. 2004. Daybreak: A scholarly biography of Canadian counsellor educator Dr. R. Vance Peavy. *International Journal for the Advancement of Counselling* 26, 177–89.

Larsen, D.J., W. Edey, and L. Lemay. 2007. Understanding the role of hope in counselling: Exploring the intentional uses of hope. *Counselling Psychology Quarterly* 20, 401–16.

Magnuson, S., L.L. Black, and M.K.E. Lahman. 2006. The 2000 cohort of new assistant professors of counselor education: Year 3. *Counselor Education and Supervision* 45, 162–79.

Menges, R. 1996. Experiences of newly hired faculty. In L. Richlin (Ed.). *To improve the academy* (v. 1, pp. 169–82). Stillwater, OK: New Forums Press and the Professional and Organizational Development Network in Higher Education.

Ochberg, R.L. 1988. Life stories and psychosocial construction of careers. *Journal of Personality* 56, 173–204.

Olsen, D. 1992. Interviews with exiting faculty: Why do they leave? In D.H. Wulff and J.D. Nyquist (Eds.). *To improve the academy* (v. 11, pp. 35–47). Stillwater, OK: New Forums Press and the Professional and Organizational Development Network in Higher Education.

Olson, M.R., and C.J. Craig. 2001. Opportunities and challenges in the development of teachers' knowledge: The development of narrative authority through knowledge communities. *Teaching and teacher education* 17, 667–84.

Palmer, P. 1998. *The courage to teach: Exploring the inner landscape of a teacher's life.* San Francisco: Jossey-Bass.

Parse, R.R. 1999. *Hope: An international human becoming perspective.* London: Jones and Bartlett Publishers.

Snyder, C.R., D.B. Feldman, H.S. Shorey, and K.L. Rand. 2002. Hopeful choices: A school counselor's guide to hope theory. *Professional School Counseling* 5, 298–307.

Turner, J.L., and R. Boice. 1987. Starting at the beginning: The concerns and needs of new faculty. In J. Kurfiss, L. Hilsen, L. Martensen, and R. Wadsworth (Eds.). *To improve the academy* (v. 6, pp. 41–47). Stillwater, OK: New Forums Press and the Professional and Organizational Development Network in Higher Education.

Van Der Bogert, V. 1991. Starting out: Experiences of new faculty at a teaching university. In K.J. Zahorski (Ed.). *To improve the academy* (v. 10, pp. 63–81). Stillwater, OK: New Forums Press and the Professional and Organizational Development Network in Higher Education.

Wisniewski, R. 2000. The averted gaze. *Anthropology and Education Quarterly* 31, 5–24.

CHAPTER 11

Finding the Time and Space to Write: Some Stories from Canadian Teacher Educators

Dianne M. Miller

I accepted the early retirement package and I did that knowing that I would go out without my promotion. So I would never be professor emeritus, even though I was told repeatedly that I was the top teacher in the department year after year. I would never get a chance to do some of the research I wanted to do but I would get a chance to write though. So, writing was one of the reasons I retired. (Margaret)

Now, I never imagined myself having writer's block because I cared about my work tremendously. It inspired me so I don't know what I thought writer's block was but I didn't think I had it. (Mary)

And my writing—it is funny now because I just feel totally blocked in my writing. I am really struggling, continually struggle with it. (Hannah)[1]

Acker (2003) emphasizes the key importance of time and place in understanding the diverse experiences of Canadian teacher educators: individual stories are embedded in shifting institutional contexts and macro-level politics such that career experiences from one generation to another and from one institution to another might vary dramatically, yet be similarly affected by broad policy trends. Study of the multilayering of micro and macro influences is intrinsic to the work of sociologists and historians such as the team who analyzed "traditions and transitions in teacher education" in three Canadian provinces. This project charted major institutional and policy transitions in teacher education since 1945 in Quebec, Ontario, and Saskatchewan. Team members were particularly interested in tracing the rise of a research culture in teacher education institutions as evidenced by an increased orientation to funded research and publication as the measure of productivity, along with higher entry-level qualifications and a more rigorous tenure process. We were concerned to document the impact on faculty members from their perspectives and in their own words (Acker 2000).

Similar research was also conducted in Iceland, Sweden, and Switzerland. Reports from the projects acknowledged increased pressure for faculty to acquire research grants and publish as well as to meet higher standards for hiring, tenure, and promotion. Increased anxiety on the part of faculty about successfully negotiating these career stages and meeting changing expectations was evident in all regions (Acker and Weiner 2003).

Drawing on data from Canada, this chapter focuses on a different perspective of time and place: the complex negotiations that individual faculty members make to find the time and space/place to engage in one aspect of the research process—to write. In particular, I am interested in those faculty members who commented on the struggle to produce work that satisfies their own creative, intellectual, and vocational impulses in an institutional context that reputedly values research and provides some mechanisms for its support; yet where support is perceived as insufficient, nonexistent, or even counterproductive for some to achieve their writing goals. I am particularly drawn to narratives of how individuals navigate time and space constraints so they can get on with the work that matters to them. These are also stories of striving for integrity in the face of challenging circumstances and, in spite of the heavy personal costs, they bear witness to and engender a sense of possibility and hope.

My personal interest in the "Traditions and Transitions" project stems from a fascination with how people understand their work as teacher educators, a work that by choice and circumstance is also mine. I began graduate studies to indulge a love for reading, thinking, and writing and found that these aspects of my life were the most unattended once I had an academic position. How do other teacher educators imagine themselves and the work they do? How do they make sense of competing demands of teaching, service to the field, and research? How do they model being the "teacher scholars" or "reflective practitioners" expected of professors and teacher candidates (Boyer 1990; Schon 1983)? Do they find a way to actually read the journals to which they subscribe or, like me, just add them to the "maybe someday" pile or let their subscriptions lapse? Does their research satisfy their own curiosity or sense of purpose or is it mostly instrumental? Do some faculty sustain emancipatory, democratic, and creative ideals in the face of neoliberal market values that increasingly bear on the mission, mandate, and meaning of postsecondary and professional education (Axelrod 1982; Dillabough and Acker 2002; McMurtry 1991)? And if so, how do they do it? Not all these questions are addressed directly in the research.

From the twenty-five or so interviews I conducted with teacher educators, I focus here on three broad overlapping themes that reveal how they articulated their struggles and the ways that they met them. The themes are: (1) A vital personal connection to research; (2) Dilemmas of time and space (including disciplinary space) to research and write; and (3) Stories of (more or less) successful navigation. Hopes for self, career, and service animate teacher educators' stories of struggle for time and space to write.

A Vital Personal Connection to Research

The Traditions and Transitions team did not make evaluative judgment about the nature or kinds of research undertaken by faculty. We are struck, nonetheless, by the many cases where research was strongly related to a sense of personal vocation and identity. Mary, in somewhat exceptional circumstances, had been hired without a Ph.D. Completion of a doctoral degree was a condition of tenure. She comments that she chose the most difficult route to accomplish this, choosing the Ivy League university with the most required courses and residency period, and fashioning her own program there while co-parenting a young family, contending with illness, and doing extra academic work to compensate for perceived lacks in her background preparation. Returning to her university of employment after an educational leave, she negotiated a further six months of unpaid leave under the "delusion" she could complete her dissertation:

> I did work on my research, but I felt like I was not happy with it. I didn't feel confident in my theoretical framework for my research. And I just felt like I couldn't fake it. That I had to own it in a way, that at that time I didn't. And I could have produced some document that probably would have been accepted as a dissertation, but I couldn't live with that. I had to feel like I had engaged in a process of learning and discovery and authentic learning.

Another important aspect of a vital personal connection to research is the eschewal of work primarily oriented toward career advancement. In the context of a strong argument for integrity in research, one faculty person expresses concern that expediency appears to rule over rigor:

> It terrifies me when I look at some of the quality of research and how poor it is...You wonder about the whole university—what the university has become and then the implications of that for the community at large. What do we have as knowledge and what do we have as wisdom? Do we have something that is true knowledge or do we have something that is fudged so that someone could get their promotion? That really worries me because it happens...because of these structures for promotion and tenure, people often choose the least ethical route (and certainly I have seen that happen time and time again) and then we get material which does not have validity. "What purpose does it have?" I guess is my question. (Margaret)

Dilemmas of Time and Space

Research is Expected but not Acknowledged as Requiring Time/Space

Mary notes that although the completion of her doctorate was a condition of her tenure, it was viewed by administrators as something apart from her work at the university. One even asked her if her dissertation was getting in the way of her work.

> I mean, at the time that this was said to me, I said in response, "I thought my dissertation was my work." And then I was told, "well, I am not telling you that you're not doing your job." And I was like, "You just did..."

A senior administrator expresses a similar frustration. One discusses being censured for taking about a month away from the university during which s/he wrote a paper:

> I mean I had started working on the paper and collecting the data much sooner than that. But it's over that time frame where I was finishing the research, doing the analysis, writing the paper. I came back, I sent the paper off...and then I go for review as [administrative position] and one of the comments was what kind of example is this to have the...gone

from December whatever to January 2nd, the busiest time of the year (which is bullshit because it is not the busiest time of the year). But nobody gives you credit for that—that I actually got a chapter written for a book in that time frame. You know people don't know what you do but again, the assumption is [when you are away] that you aren't doing anything. (Jane)

Dilemmas Created by Program Restructuring

Shortly after she was hired, Hannah's department completely reorganized its programs in response to an external review. She found herself participating in program revisions that erased her own area of expertise. At the same time, an expectation that faculty research would support the department's new focus was made clear. This meant a radical shift in Hannah's research orientation that she likens to adjustment after amputation:

> It is not as though good work hadn't been done in the department before, it is just that things have changed and it was time to review and rejuvenate and up-date maybe. But, in the course of doing that, I don't feel like all faculty resources were taken into account. I don't know if what they could best offer or most capably offer—So, I mean I sort of feel like one of my arms got cut off. So, where I used to have the ability to use both my arms, now it is probably my left arm (like my good arm is gone). So, it is stressful in that way. (Hannah)

She describes the impact on her research:

> So, research, what I ended up trying to do has really pulled me away from where I was going—so I try to marry some of the interests that I have with more of [department's] focus. And, to be perfectly honest, I feel it has gotten me into a little bit of trouble. In trying to resolve, trying to make something fit in a way that I could make it work for me with what are overwhelming demands here...it is a whole new area [of research] that I am immersed in. (Hannah)

A colleague from the same department reflected a similar sentiment about not fitting in and questioned whether there would be a place for his research interests in the new focus:

> ...what's happening is that we are developing a new research culture in the department and it is not necessarily a culture that I share. I

> keep on getting the message that "well, we have a place for you." So that's nice that you have a place for me but you know, can I control the place? (Ted)

Stories of Navigation around Time and Space Constraints

The navigational strategies used by teacher educators to find time and space to write resonate with poet Emily Dickinson's (nd) phrase, "success in circuit lies."

Draining the Swamp, Digging New Holes

Several interviewees attest to a variety of work pressures. Some share a sense of exhilaration and adventure in responding to these pressures, acknowledging that faculty positions in universities still offer enormous opportunities for creativity and purposeful work.

> That's the beauty of this job: you make it what you want to make it. And you can bitch like I was bitching about too much work, but if you sit down and analyze what I am doing...correct. I dug the damn hole. I poured the water in. I threw the plants in. I threw the crocodiles in, then I jumped in and now I am saying "Shit." But that's the beauty of the job. I can drain the swamp and dig another hole. I don't know about you, but I like that. (Benjamin)

Others feel frustrated and diminished. Margaret retired at the "top of her game" as far as teaching went because she wanted to leave before she had trouble "finding the door." But she also wanted to leave the "angst and the attacks and the terrible atmosphere" and retirement seemed the only available recourse. Her postretirement writing focuses on the importance of research in determining and supporting deep level change:

> ...what I try to bring out in what I wrote is that it is so hard for teachers to distinguish between what is simply administration initiative in that school or in that school division, and what is a deep level of change. Some things are just surface level things but when we come to deep level change, it is a commitment and it is hard. And, as a profession [we need] to find the energy to stop the bombardment, do the reflection, and do some deep level change. Because as long as people are bombarded—we talked about my research, you finding time to do your research—as long as people are bombarded with tasks and meetings (meaningless meetings

often)—this deeper level change and the advancement of what we do in our profession I think is jeopardized... there are ways to honor those who have had time to do the research so that we incorporate that into our practice. And I think often we don't do that. We don't leave that and say okay that obviously is solid research. We need to take that into account. We need to model that...

Getting a Room of One's Own and a Coach

Mary worked to resolve the dilemma of time and space to write by limiting her time spent at the university and renting office space at her own expense to get away, at least temporarily, from family and other work obligations:

> After a while in the [university] job, I began to realize that if I didn't get the dissertation done then I wouldn't have a job there anyway. And so, do what it takes and if that means not being at work 'til the afternoon, that's what I did. So, I had my teaching schedule changed to the afternoon. So beginning in the fall of 200x, I didn't come to work in the mornings anymore because I got my own office away from campus—no phone, no email. So these were the things that I needed, that I had to do.

Mary also credits her discovery of a Web site and teleconference workshops for ABDs (All But Dissertation) for making it possible for her to get her dissertation done.

> But when I heard that teleconference conversation, and that evening 150 ABDs around North America had called in... Well I was blown away because I was sure I was the only one struggling in all of North America. So that, in itself, and the fact that what he [leader] talked about—what he went through—I was going through. All the self-doubt, all the—there was just, I cried when I heard that, when I listened to that workshop. And then I knew that I wasn't alone. And then I knew that I could get help and I did.

Based on advice received from this Web site, Mary hired a coach to whom she reported weekly. The coach did not usually read her work or provide feedback: "the deal was that she just be this coach and mentor; someone to be accountable to." With the support of the coach and teleconferencing workshops, "it [the writing] started to happen right away."

Lifting the Thousand Layers to the Soul

Previous to her employment in the university, Hannah had worked with terminally ill patients in a hospital. The realization that we all have a limited time to live—that we are all dying as we live and still living as we die—helped her develop an appreciation for being as "gentle as you know how to be" because "these are other souls you are with." It taught her the importance of honoring the spirit in the person. This became a touchstone for her in the university environment, where "I don't think that we try intentionally to live that way very often." People who know that they soon will die become very clear about what matters to them, but in the university there is often "a thousand layers to the soul." The insights she gained from her former patients gave her a clear perspective about her work at the university:

> I decided that whatever I was or wasn't here, or did or didn't do, or succeeded at or didn't succeed at, that [honoring the spirit] is sort of like a grounding point for me. That's actually more important to me than any paper I may or may not write or research I might or might not do—somehow if you stay connected to that, it's probably possible to do your work more easily.

Conclusion

Highly dedicated professors often are challenged by the contexts in which they work, to find the time and space required to conduct research and to write, especially to write. Academic environments are usually characterized by heavy and competing demands, a chilly collegial climate,[2] and an increasingly singular focus on research as the measure of productive work. Some teacher educators who express a vital personal connection to research go to extraordinary lengths—renting writing space, limiting their time to other tasks, retiring—to accomplish their goals. While some solutions may seem extreme, the stories of these educators are also infused with a sense of commitment, purpose, and hope. They speak the language of the heart (Apps 1996; Palmer 1998; Intrator 2002). Their stories remind readers that, while many forces are brought to bear on life in the academy that are not within an individual's control, university educators can take responsibility to participate in and develop the professional culture in 'good' ways—ways that affirm intellectual and spiritual growth, meaningful research, good relations, and that foster a sense of hope for a better world.

Notes

1. These quotations and others identified by name (pseudonyms) throughout the text are taken from the Social Science and Humanities Research Council of Canada (SSHRCC) study, "Traditions and Transitions in Teacher Education: The Experiences of Teacher Educators Ontario, Quebec, and Saskatchewan, 1945–2002," by Sandra Acker (Principal Investigator), Jo-Anne Dillabough, Dianne Hallman, Therese Hamel, and Elizabeth Smyth.
2. References to "chilly climate," particularly as experienced by women in the academy, are abundant; see Freyd and Johnson 2008.

References

Acker, S. 2000. Traditions and transitions in teacher education: The development of a research project. *Historical Studies in Education* 12(1–2), 143–54.

———. 2003. Canadian teacher educators in time and place. *Tidskrift för lärarutbildning och forskning/Journal of Research in Teacher Education* 10(3–4), 69–86. Retrieved January 31, 2009, from http://www.educ.umu.se/presentation/publikationer/lof/lofu_nr3-4_2003.pdf

Acker, S., and G. Weiner (Eds.). 2003. *Tidskrift för lärarutbildning och forskning/ Journal of Research in Teacher Education* 10(3–4). Retrieved January 31, 2009, from http://www.educ.umu.se/presentation/publikationer/lof/lofu_nr3-4_2003.pdf

Apps, J. 1996. *Teaching from the heart*. Florida: Kreiger Publishing.

Axelrod, P. 1982. *Scholars and dollars: Politics, economics, and the universities of Ontario (1945–1980)*. Toronto: University of Toronto Press.

Boyer, E. 1990. *Scholarship reconsidered: Priorities of the professoriate*. Princeton: Princeton University Press.

Dickinson, Emily. nd. "Tell all the truth but tell it slant—". 1129 *Wikisource*. Retrieved June 28, 2008, from http://en.wikisource.org/wiki/Tell_all_the_Truth_but_tell_it_slant_--

Dillabough, J., and S. Acker. 2002. Globalization, women's work, and teacher education: A cross-national analysis. *International Studies in Sociology of Education* 12(3), 227–60.

Freyd, J., and J.Q. Johnson. 2008. References on chilly climate for women faculty in academe. Retrieved June 28, 2008, from http://dynamic.uoregon.edu/~jjf/chillyclimate.html

Intrator, S. (Ed.). 2002. *Stories of the courage to teach: Honoring the teacher's heart*. San Francisco: Jossey-Bass.

McMurtry, J. 1991. Education and the market model. *Journal of the Philosophy of Education* 25(2), 209–18.

Palmer, P.J. 1998. *The courage to teach: Exploring the inner landscape of a teacher's life*. San Francisco: Jossey-Bass.

Schon, D. 1983. *The reflective practitioner*. New York: Basic Books.

CHAPTER 12

Learning about Hope through Hope: Reflections on the ESL Enterprise

Judy Sillito

I marvel at the audacity of hope. It is this audacious nature—the spirited, daring, boldness—of hope that creates surprising outcomes to disconcerting circumstances that seem inevitably doomed. I stand in awe before the hope that I witness daily in the struggles of the immigrant learners I teach; I long to give voice to their stories; and I am moved, inspired, and propelled by the hope that orients my spirit. We live storied lives and my life has been storied by hope. Thus, I am called to understand what it means to hope—as individuals, as learners, and as teachers. And so begins the story…

What Brought Me to Hope

For over a decade I had been happily, and I might say successfully, working as an English as a Second Language (ESL) teacher to immigrant and refugee adults. Suddenly my world was turned upside down when I became a widow with four young children. I withdrew from the world of work for a time and when I was ready to resume, I found that I could no longer face the suffering, the deep loss, that I was witness to in my classroom. In the eyes of each refugee and immigrant, persons who had been displaced and disrupted, often worse, I recognized my own loss, and connected with theirs. But the pain was still far too close.

During the next five years I stayed connected to the field as a program coordinator and it was then I was first introduced to the Hope Foundation of Alberta. I was instantly captivated because I was now able to name my experience. I recognized it was in the depth of my despair that I first knew my hope. I realized then how much I was relying on the hope that emanated so freely from the hearts of children, as my little ones led me down grief's wondrous and painful path.

Diving into the vast library on hope literature at the Hope Foundation, I was struck by an excerpt written by founder Ronna Jevne (1991). In response to an incredulous query as to why anyone would do a job (working in a cancer hospital) that is privy to so much suffering, her reply was

> to the outsider I'm sure it's disconcerting... but you don't see the courage, the hope... You don't see the dedication and vision of an incredible team doing an impossible task... I can think of very few people who in their work have the privilege of seeing courage every day. (Jevne 1991, p. 11)

She, of course, was referring to her work in palliative care, but she could have been speaking about the classroom of refugees and immigrants I found myself standing before.

It was in that instant that I became determined to learn more about hope and so entered graduate school to eventually use narrative inquiry methods to research hope and transformation with adult ESL learners and teachers.

The Study of Hope

There are many models that try to pin down the elusive, intangible qualities of hope. Dufault and Martocchio's (1985) multidimensional model seems to encompass the essence of the ongoing scholarly discussion. Although one-dimensional frameworks exist (Snyder 1994; Stotland 1969), hope is more readily understood as consisting of multiple dimensions including cognitive, relational, affective, contextual, behavioral, and temporal spheres (Dufault and Martocchio 1985) and being a rational, experiential, and transcendent process that is irreducibly linked to hopelessness (Farran, Herth, and Popovich 1995; Hafen et al. 1996). Hope is related to one's sense of purpose, and involves factors of risk and authentic caring (Nekolaichuk, Jevne, and Maguire 1999) and its depictions are often most easily shared through stories (Edey 2000; Jevne 1994).

I personally find no dissonance with such academic descriptions of hope and am thrilled that there is growing interest in the study of a phenomenon that historically has been broached with skepticism from an academic community thinking it too vague to be worthy of investigation. However, there is one exquisite definition that sums up both my intimate discovery of hope and my reflections upon this research. This definition comes not from a hope researcher but from a playwright, a prisoner, a political exile, a president:

> Hope is an orientation of the spirit, an orientation of the heart, it transcends the world that is immediately experienced, and is anchored somewhere beyond its horizon... It is not the conviction that something will turn out well, but the certainty that something makes sense, regardless of how it turns out. (Havel 1986)

In considering the balance of the theoretical and more spiritual expressions of hope, I designed a research project to look deeper into the way immigrant and refugee learners interpreted their hope, looking for clues on how this might broaden our Western understanding and at the same time enhance the successful integration of newcomers to Canada through a teaching methodology that openly invited hope. The following is a description of the study that evolved.

This qualitative study explored the narratives of five ESL learners and two ESL teachers (Sillito 2004). An initial focus group opened the hope conversations after which participants were invited to share their stories through a variety of narrative inquiry methods including interviews, journals, photographs, and orally told stories. Each participant was given a journal and a disposable camera to record their reflections on hope. They each participated in the initial focus group as well as in two interviews over the course of three months. The research was conducted with the overarching question "What do adult ESL learners and teachers say about their notions of hope, with particular reference to the relationship of hope to their learning experiences?" A series of guiding questions was used to shape the inquiry of the reflections, stories, and images: What are the critical incidents that enhanced, diminished, or made visible the hope of this sample group? Is there any discernible relationship between the hope of teachers and the hope of learners? How can an understanding of hope inform our understanding of the transformative learning process? Can hope itself be an emancipatory process and, if so, how aware of the emancipatory nature of the hoping process are those that hope?

In this chapter you will read excerpts about hope from the research participants, all of whom have been assigned pseudonyms. You will be able to hear, in their own words, thoughts about hope from Maria (Sierra Leone), Hong (Cambodia), Yang (Hong Kong), Sarim (Iran), David (Romania), and two Canadian teachers, Janet and Angela (Sillito 2004).

Narrative research design is a qualitative approach in which the researcher collects stories generated by the participants and narrates the experience in rich detail (Creswell 2002). When retelling the stories, the research concentrates on specific situations or ideas and analyzes them for themes. Existing research into the nature and experience of hope supports the usefulness of storytelling (Edey 2000; Jevne 1994).

When asked to describe what hope means, people often choose to tell a story to illustrate their meaning (Edey 2000; Jevne et al. 1999). This would suggest we can understand knowledge of hope in much the same way as Clandinin and Connelly (1998) understood teacher knowledge—as stories lived, told, and retold. The value of narrative inquiry is enhanced when we "go beyond the simple recounting of a story to a more educative practice of retelling and reliving stories with imaginative possibilities" (Whelan et al. 2001, p. 1). This echoes the approach prescribed in hope-focused counseling (Edey 2000) where the listener makes a conscious choice to listen to and engage with the teller in imagining hopeful ways of retelling a story. By using the language of hope (Edey 2000), the listener helps to make visible the hopeful possibilities in the stories. Hope is enhanced through the exchange and the teller can relive the story from this new perspective.

I did not begin the research journey with a hypothesis to prove. I began with an openness to listen to what others told me about hope, and an openness to uncover those things that I had no way of knowing that I did not know.

What the Study Revealed

Many of the essential meanings of hope expressed in the findings of this research echo the attributes of hope described in the literature. Although there is no consensus in the definition of hope, nor does there need to be (Jevne 1991), some attributes occur with significant frequency. This suggests that there may be universals in the concept of hope that are experienced in similar ways across cultures, genders, languages, and worldviews.

The stories of hope in this study center around four themes, which I have named the existential, relational, rational, and transcendent "natures" of hope. In organizing the data around the natures of hope I propose a conceptual framework that exposes the personal meaning of hope and honors the uniqueness of human experience; in that way, I hope to offer a tangible framework for the intangible experience of hope.

The *Existential Nature of Hope* refers to the personally meaningful hope that "... just is" (Janet). It is the kind of hope of which Havel (1986) speaks—a life force that "has always been there starting from day one" (Hong) and "is like an ever-present pilot light" (Janet). Hope as a life force is a characteristic that reiterates throughout the literature. Dufault and Martocchio's (1985) definition of hope refers directly to hope as a life force. Others (Farran et al. 1995; Hafen et al. 1996; McGee 1984; Nekolaichuk et al. 1999) describe hope as a motivating force or necessary condition for action. The results of this study lend credence to this understanding of hope. The idea that hope is a necessary life energy and that without hope there could be no life is a theme that emerged to greater or lesser degrees in the responses of all the participants.

I was prompted to consider possible connections between humor and the transcendent or existential nature of hope as I listened to the stories and shared moments of laughter. However, the idea that humor may be one avenue by which to engage the transcendent is not broached in the literature. In Dufault and Martocchio's (1985) model, humor would seem to place neatly in the affective domain, whereas humor in relation to transcendence would fall under what they call the affinitive domain. Nekolaichuk et al. (1999) refer to the state of happiness under the factor of authentic caring in the interpersonal dimension of hope, but the profundity of humor in the stories of these participants seem to fit better across the entire scope of attributes of personal spirit, such as vibrancy, meaning, valuing, engaging, caring. Further research is obviously needed to investigate the possible relationship between humor and the transcendent. Nonetheless, hope literature and this data support the importance of humor in hope.

The *Relational Nature of Hope* describes the propensity of hope to be experienced within relationship and is also well documented in the literature (Dufault and Martocchio 1985; Edey 2000; Edey et al. 1998; Farran et al. 1995; Nekolaichuk et al. 1999). Hope is both enhanced and diminished within relationships. However, this narrative study offered new texture to the meaning of relationship that is not forthcoming in the literature. Throughout the literature, and so too in this study, hope is often witnessed within relationships between people—family, friends, lovers, strangers; yet these narratives also include relationships with community, culture, a supreme being, and nature. There is also a focus on the sharing of humor as a powerful force that connects people to one another and to their hope. Only Edey (2000, 2003) and Jevne (1993) discuss humor to any great length. Our findings support their acknowledgment of the importance of humor in the landscape of hope and call to the further exploration of the connecting power of humor. Service to others constitutes another important aspect of connectedness to one another. It sustains and elevates hope. In all these variances of human interrelationship, hope is more easily accessed, more readily experienced, more dramatically impacted when it is in relation to others.

The *Realistic Nature of Hope* is authentic hope. It is this realism that separates hope from optimism. It is not necessarily the goal of hope to be realistic; rather, being realistic is the product of being genuinely hopeful (Faran et al. 1995; McGee 1984; Snyder 1995).

A goal or desired outcome is an essential element of hope that has been addressed from several different perspectives in the literature (Edey and Jevne 2003; Dufault and Martocchio 1985; Farran et al. 1995; Snyder 1995; Stotland 1969). The findings in this study strengthen the position held by many that goals are not rigid or static. They can change over time, or shift within the hoping experience as a person gradually relinquishes hopes that become unrealistic or unattainable and replaces them with a focus on new hopes.

Dufault and Martocchio (1985) describe two kinds of hope, particularized hope and generalized hope. Generalized hope describes hope that has no specific object, a general feeling of hopefulness that keeps one going. Particularized hope, on the other hand, is focused on a hope-object. A person's orientation between the two kinds of hopes vacillates as particularized hopes are identified and threatened. Dufault and Martocchio (1985) suggest that

> particularized hope clarifies, prioritizes, and affirms what a hoping person perceives as important in life, [whereas] generalized hope provides the climate for developing particular hopes and later rescues the hoping person when the particular hope no longer seems realistic. (p. 381)

Clearly in the stories of refugee camps and escapes there was an umbrella of hope that propelled some through the perilous times of uncertainty when particular goals were untenable. The voices in these narratives affirmed that at times there is nothing left to do but hope and "hope is quite enough" (Dufault and Martocchio 1985, p. 381). The notion of societal hope, and a hope for world peace, although they have objects, also seem to lean more toward a generalized hope that buoys the spirit when particularized hope is threatened.

The *Transformative Nature of Hope* speaks to the irreducible link between hope and change. The word change and the theme of growth, learning, and a yearning for development figure very strongly in the stories. Change equals hope, and hopelessness is associated with either no change or fear of change. Not all change indicates that transformative learning has occurred. But all transformative education requires that something change. This study bears testimony that hope is part of that dialogue of change.

Aside from a bid to deepen our theoretical understanding of hope, a second dimension to this research project looked at the specific relationship between hope and the ESL education enterprise. Four themes emerge from the narratives and within the themes there is at times, agreement in the views of the teachers and learners. Sometimes, however, the perspectives were quite different. In fact, at times, their hopes conflicted. The following is a summary of the salient themes.

Education is Hopeful

Most emphatically for the learner group, education was seen as very important in life and with regards to hope. For each of them, the hopes they held for coming to Canada were related to a better education for themselves and/or their children. Even in the case of a pharmacist who had to sacrifice his own education to support his family, education was seen as hopeful and very important. Although employment was a prime goal of education, the learner participants felt education itself elevated their hope. Yang studied while employed and continued to study after her retirement saying, "I am happy now because I can go to school. In Hong Kong impossible to study when you are old." Hong spoke of his diminished sense of hope when his family discouraged him from pursuing another career because it wouldn't lead to a higher-paying job. Maria said simply, "I have hope in my school."

Curiously, the narratives of the two teacher participants did not speak at all about the hopefulness inherent in education.

Language is Power

Here, teachers and learners shared a common understanding. Both sides understood the purchase of the English language in a globalized world. The learners talked about their enhanced power as a result of being able to get around in the new language. Yang said, "Many countries speak English. Many countries have foreigners so English is the first language in the world. Study English is important because this convenience myself." Another learner discussed his consternation with his accent, feeling that it set him up for discrimination. One of the teacher participants referred to "how language becomes a vehicle for hope...I find our language and how we teach it carries a message of hope in a way...you have some control once you get a handle on your communication skills."

Teachers Should Listen

Teacher and learner participants were unanimous in their opinions about listening and hope. Listening is seen by the learners and teachers alike as the most important quality for an ESL teacher to possess. Both groups value a learning environment that is predicated on a good relationship between teacher and learner. This relationship includes humor and respect, but above all listening. The teachers feel their own sense of hope supported through actively and attentively listening to their learners, in order to identify their needs. Conversely the biggest strain on their hope, in the case of both teacher participants, is experienced when they cannot seem to figure out what a learner needs and the learner fails to progress well in the class. Learners repeatedly recalled stories of how their hope was elevated by a teacher just listening. The teachers in their stories did not offer any profound advice or do anything spectacular. They simply listened—for a long time. And this alone elevated the hope of the learners. Teachers appreciate programs that allow the kind of flexibility and autonomy that enables them to listen and respond to the needs of their students.

Marks Matter

Related to being responsive to the individual needs of the learners was the teachers' tension with marks. Teacher participants saw grades as not being reflective of the value of a learner's work. They were more interested in the learning process, improvement, and change that could be observed in the learners. They attested to the subjective and rather arbitrary nature of grading, especially in the context of language learning and cultural integration. Another factor that both teachers discussed is that ESL classes are often non-credentialed courses, so the marks are largely irrelevant in the larger context of education. Angela says,

> Just relax. How important is this grade I give you rather arbitrarily. We each mark differently. Do you really think the number I give is reflective of the person you are? And I hope they resize that they are more important than whatever number I give them. How I grade them is so different than you would grade them or another teacher. I don't want them to take it so seriously.

However, when learner participants were asked to relate a time when their hope was challenged in their ESL classes, the stories in every case revolved around marks; either marking they considered unfair, or getting poor marks. In the same vein, good marks and academic promotion figure prominently in the stories of hope in the learning context, second only to being listened to by their teachers. This cautions teachers to be wary of ways in which they may inadvertently diminish the hope of students by issuing low grades that they believe matter little in the narrative of life. This is not to advocate for meaningless high marks for all learners; it does however challenge ESL educators to reflect on the relevance of traditional mark-based evaluation systems in the context of adult ESL. Personally, this finding has caused me to reassess my own teaching and evaluation methods.

Insights

This project was a study about hope, through hope. The participants told their stories about hope, and it was my intention to hear and probe their stories with a hopeful presence. I chose to view our encounters through the lens of hope as expressed by these seven participants. As a result, the happenings within the research process itself shed as much light on the way hope works as the actual responses of the participants.

Insights from the research support Clandinin and Connelly (2000) who contend that, because of the nature of narrative, it is important to have strategies to recognize and assess the multiple levels at which inquiry proceeds. Indeed, in this study, there were many examples of layered narratives whose deepest meanings were told through the telling and retelling, listening and reflecting on both the content and the process of narratives.

One such example comes from Janet, a teacher participant who entered the research endeavor in the middle of marital crisis. As Janet pondered her experience of hope and hopelessness, she explained, "I get down but I don't dwell there. Hope lets you do that. It lets you visit some yucky places because it's kind of your lifeline that brings you back."

She credits the dialogue about hope and her hope journal with enabling her to enter counseling and be strong and true in the process. In her words, "These two things [successful marriage counseling and hope journaling] are serendipitous and I don't think co-incidental." She also was able to reflect on the social forces that were serving to oppress her control, and dampen her hope. As an example, she compared her two journals, the one she kept as a research participant, and the one the marriage counselor had suggested she keep as a way of venting her thoughts. Accepting the assumption that to vent her anger on paper would be therapeutic, Janet wrote pages and pages of what she later called "poison." She explained how the focus on her husband and on her rage festered, fostering increased rage and diminishing her ability to act. She could not get past the rage enough to even imagine a way to escape her situation; much less have the will and the ways to enact change.

At the same time, I had asked Janet to keep a hope journal as part of this research exercise. She did so, and remarked being surprised to notice the profound effect her hope journal had had on her. Although she was not intentionally writing about her marital situation, in the process of focusing on hope, she found herself newly aware of choices and possibilities that were open to her within her marriage, and newly refreshed with the energy to put those changes into motion. Unlike the venting journal, the hope journal reminded her of other times when she had hope and used hope, and she says:

> I'm not sure I would have been so open to the counseling had I not participated in these hope conversations. And it has given me power. And power is hopeful. The power to reassess. I am certainly feeling more hopeful and powerful in my marriage. I really think hope let that open up. Definitely.

She also reported being much more confident about her ability to change without compromising what was important to her as a woman, a wife, and as a mother. Her narrative was altered by the way she storied her world. Writing "to vent" reinforced her story of victimization, while writing through a lens of hope reframed her story and thereby transformed her lived narrative.

The literature is clear that an intentional focus on hope helps foster hope and an apt instrument to draw such focus is the narrative. This study demonstrates the power of articulating a story, whether through a relationship of telling and listening, by a process of writing and discovery, or with metaphors explained through photographs. For Hong, exploring his narrative orally prompted him to put pen to paper, something he had not done for years. He explains:

> I think when I go back home I'm going to start writing. Thank-you, thank-you for this opportunity. I never sat down to think about this, but I'm going to and I'm going to write it so I could use it one day. You know in Cambodia people, many people have nothing. Some of them believe they have nothing but I want to tell them they have hope. So I have to write this down so I could apply it to every situation. It's really good. Because when you bring hope to others you fulfill your own hope.

Sarim simply said, "Things have changed so much for me, this talking about hope."

One woman, however, offered a disconcerting response that tells of the more mysterious layers of narrative inquiry.

Maria was exuberant and full of stories during our first interview. She talked freely and readily for over two hours. We left one another in good cheer and she said she would be happy to do a second interview. We made an appointment for late the following week.

The second interview was vastly different in tone from the first. At the outset I noticed Maria seemed tired so I asked if she wanted to reschedule. She declined and we began. I started to ask some questions of her photographs and immediately she seemed exasperated. She talked, without her usual animation, about hope in the abstract—lofty words bereft of the stories that were so abundant in the first interview. Later, the stories resumed, and they were very sad and personal accounts of despair. I tried to be sensitive to her emotion, listening attentively without pressuring her to continue talking about that which she would rather forget. But with every comment I made to create an opening for her to change the subject if she desired, she reverted back to the next horror story. Her body language changed, too; she began to cower. My dis-ease was also growing and I did my best to alleviate our distress, to little avail. Maria's answer to the last interview question (What was it like for you to participate in this research?) reflected the contradictions that had dominated this interview:

> I just enjoy it because you make me remember things I want to forget and you make me remember the important of hopes, hope. Anyway it is okay, I enjoyed it. It's great.

I wondered why and how remembering horrors that she would rather forget could be enjoyable, great, or even okay; especially when her general demeanor was indicating that it was not okay at all.

Perhaps my attempt to be overt and academic about something as sacred as hope was, for this woman, an ineffective approach. For Maria, whether for reasons of culture, race, ethnicity, religion, spiritual paradigm, or perhaps a soul branded by torture, analyzing hope in this critical way may have been insulting, patronizing, and demeaning, in addition to being ineffective. I noticed whenever I asked her to talk about her photographs she got a quizzical look on her face that was full of discomfort. She was not bubbling to talk about her pictures as the others were, but instead would say "Just look and you will see." She seemed perplexed, perhaps offended, that their relevance required explanation.

Ethically, I wondered if I had transgressed. Analytically, I wondered where the power was and who was controlling the interview. Spiritually, I wondered where my hope was going. Reflection of my own position in this interview shed light on the workings of narrative inquiry.

When research focuses on the primacy of experience, respect for the beliefs and values of others, an emphasis on relationship and process factors, and a search for authenticity, the relationship between researcher and participant "renders the respondent vulnerable and the researcher responsible"

(Hart 1999). In this study, the rules of engagement were blurred, given the very personal nature of both the topic and the methodology. With my tape recorder and my note pad, I could not escape the trappings of being perceived as a researcher, garbed in objectivity, seeking answers to questions for my own purposes. However, within the research relationship, I was one woman speaking to another with a level of trust and understanding that opened the path for sharing deeply personal ideas, memories, and impressions. As the contradiction emerged in the interview, I became more aware of, and uncomfortable with the tape recorder.

Weeks later, in an interview with a different participant, I encountered a similar situation, as one man became very emotional as he poured out his despair. This time I turned off the tape recorder. This was not to be fodder for my research purposes because this seemed not to be the story he had agreed to share with me when he signed on to talk about hope. This emotional outpouring actually contained little in the way of information about this individual's understanding of hope, or hopelessness for that matter. But the discourse itself gives testimony to the authentic relationships that can develop in narrative inquiry and supports the assertion made by all participants that listening is a powerful way to elevate hope. Once I turned off the tape recorder I felt immediately more centered in my ability to be one person listening to another, and less like a voyeur of anguish. The result of this decision was a much more empowering and respectful encounter.

Emancipatory research invites and facilitates avenues of change, and with the potential for deep meaning and profound change comes the potential for great exploitation. So, the emancipatory researcher must pay rigorous attention to ethical considerations.

Narratives about experience cannot be taken "as 'brute data', transparent in their meaning" (Campbell 2003). As a researcher purposely working through hope, I irrevocably enter and affect the narratives. Nonetheless, I tried to respond to the stories in the listening and the retelling by taking a stance described in Campbell (2003) as moral deference. This calls me to be true to the essence of the expression whether or not it resonates with my personal perspective. Campbell adds that the moral obligation of a listener and interpreter of narratives is to "render salient what was salient for her in the way it was salient for her" (p. 233). By having the participants respond to the images, narratives, and interpretations that I heard and retold, I learned a great deal about the accuracy, fairness, and authenticity of the data I gathered. In addition, in keeping a journal detailing an audit trail of all the interchanges and reflections, I was able to maintain a vigilant awareness about my own impact on that expression.

My Story Retold, Relived

Five years after tragedy forced me to retreat from the classroom, I returned to teaching with new stories to live by. My teaching is different. I still see the stories in the eyes of my students. I still hold their stories preciously, but the pain (theirs or mine) is no longer overwhelming. Through the research process and analysis I have concluded that my practice must allow space for three powers: hope, humor, and hearing. Hope, because the mere focus on hope elevates hope; humor because of its power to elevate hope and mitigate pain; and hearing, a concept that goes beyond listening as it involves being present to one another as whole persons and hearing messages that come from sources other than words. These are the anchors for those who dare to hope.

References

Campbell, L.A. 2003. *Discourse of agency and community: Northern women administrators.* Ph.D. diss., University of Alberta.

Clandinin, D.J., and F.M. Connelly. 1998. Personal experience methods. In N.K. Denzin and Y.S. Lincoln (Eds.). *Collecting and interpreting qualitative materials* (pp. 150–78). Thousand Oaks, CA: SAGE Publications.

———. 2000. *Narrative inquiry: Experience and story in qualitative research.* San Francisco: Jossey-Bass.

Creswell, J. 2002. *Educational research: Planning, conducting and evaluating quantitative and qualitative research.* Columbus, OH: Merrill Prentice Hall.

Dufault, K., and B.C. Martocchio. 1985. Hope: Its spheres and dimensions. *Nursing Clinics of North America* 20(2), 379–91.

Edey, W. 2000. The language of hope in counselling. Unpublished manuscript. Alberta: The Hope Foundation.

Edey, W., and R. Jevne. 2003. Hope, illness, and counselling practice: Making hope visible. *Canadian Journal of Counselling* 37(1), 44–51.

Edey, W., R. Jevne, and K. Westra. 1998. *Key elements of hope focused counselling: The art of making hope visible.* Edmonton, AB: Hope Foundation.

Farran, C.J., K.A. Herth, and J.M. Popovich. 1995. *Hope and hopelessness: Critical clinical constructs.* Thousand Oaks, CA: SAGE Publications.

Hafen, B., K. Karren, K. Frandsen, and N. Smith. 1996. *Mind/body health: The effects of attitudes, emotions, and relationships.* Boston, MA: Allyn and Bacon.

Hart, N. 1999. Research as therapy, therapy as research: Ethical dilemmas in new-paradigm research. *British Journal of Guidance & Counselling* 27(2), 205–10.

Havel, V. 1986. Public address. Retrieved August 1, 2003, from www.socialpolicy.org

Jevne, R. 1991. *It all begins with hope.* San Diego, California: LuraMedia.

———. 1993. Enhancing hope in the chronically ill. *Humane Medicine* 9(2), 121–30.

Jevne, R. 1994. *The voice of hope: Heard across the heart of life.* San Diego, CA: LuraMedia.

Jevne, R., C. Nekolaichuk, and J. Boman. 1999. *Experiments in hope.* Edmonton, AB: Hope Foundation.

McGee, R. 1984. Hope: A factor influencing crisis resolution. *Advances in Nursing Science* 6, 34–44.

Nekolaichuk, C., R. Jevne, and T. Maguire. 1999. Structuring the meaning of hope in health and illness. *Social Science & Medicine* 48, 591–605.

Sillito, J. 2004. *Images and stories of hope: Understanding hope and transformation with adult ESL learners and teachers.* M.A. Thesis, University of Alberta.

Snyder, C. 1994. *The psychology of hope: You can get there from here.* New York: Free Press.

———. 1995. Conceptualizing, measuring, and nurturing hope. *Journal of Counselling and Development* 73, 355–60.

Stotland, E. 1969. *The psychology of hope.* San Francisco: Jossey-Bass.

Whelan, K., J. Huber, C. Rose, A. Davies, and D.J. Clandinin. 2001. Telling and retelling our stories on the professional knowledge landscape. *Teachers & Teaching* 7(2), 143–57.

CHAPTER 13

A Newcomer's Hope: A Narrative Inquiry into One Teacher Educator's Professional Development Experiences in Canada

Yi Li

Introduction

A life as an immigrant is a life of unknowns and uncertainties. With high hopes of finding a better life many immigrants come to Canada. Yet they do not know about the new context or place. Figuring out new contexts and ways-of-being is a long and sometimes very painful process. In my case, I did not find my "new" life better in any sense of the word immediately after immigrating to Edmonton, Alberta, Canada. It was so different from what I had imagined before immigration that I longed to return to my home country, China, where I felt I belonged and lived a much happier life.

In order to move away from the typical "new immigrant" jobs, I began the master's program in Second Language Education at the University of Alberta in September 1998. At that time, I was hoping to discover the best method for teaching English as a foreign language (EFL) in China. I thought that I would only remain in Canada for two or three years, and would eventually return to Shanghai to teach. Prior to moving to Edmonton, I had been a university English instructor for almost ten years, three of which included teaching English content in a program for preservice English teachers in secondary schools. It never occurred to me then that I would pursue doctoral studies and that I would one day stand in front of a Canadian university classroom and teach.

Narrative Inquiry as a Theoretical Framework

In her keynote address "Creating Narrative Inquiry Spaces in Teacher Education" at the Fourth International Conference on Language Teacher Education at the University of Minnesota, Dr. Jean Clandinin (Clandinin, Steeves, and Chung 2007, pp. 18–20) proposed a narrative reflective practice approach to engage teachers in reconsidering their teaching practices. Underlying this approach are the three interwoven narrative conceptualizations: a view of teacher knowledge as experiential, embodied, emotional, moral, personal, and practical; a view of teacher knowledge as composed and recomposed as teachers live out their lives; and a view of teacher education as a possible place for sustained narrative inquiry into teachers' lives.

In their teacher education courses, Connelly and Clandinin (2006) use "living," "telling," "retelling," and "reliving" to help course participants structure and engage in a narrative reflective practice approach. People are encouraged to tell stories of their lives. These tellings are then shared with others in the class and are used to write a narrative of each person's living. Through ongoing response and people's own inquiry into their lived and told stories, they begin to retell their living, that is, to interpret their lives as told in different ways, to imagine different possibilities for their future practices. Through this retelling, this narrative inquiry process, people sometimes begin to relive their new retold stories and a shift in their teaching practices takes place.

In the pages to follow, I will narratively inquire (Clandinin and Connelly 2000; Johnson and Golombek 2002; Clandinin, Steeves, and Chung 2007) into two storied moments that depict my learning journey of becoming a second language teacher educator in Canada. I shared these two stories at the weekly meetings of Research Issues[1] at the Centre for Research for Teacher Education and Development (CRTED) at the University of Alberta in the fall of 2006. Interspersed between these two narrative tellings are my retellings of this journey. I will conclude this chapter by imagining how I might begin to relive a new retold story of my future teaching practices as a second language teacher educator in Canada.

Learning to Teach in a Canadian University: Two Narrative Moments

With a high TOEFL (Test of English as a Foreign Language) score of 630 and ten years of English teaching experience in China, I was very confident about my ability to finish the master's program in two years. Little did I know that my transition into a very different educational system would be filled with difficulty and uncertainty. Like most newcomer graduate students from other countries, I felt lost in this seemingly strange Canadian educational context. The mere thought of standing in front of a Canadian classroom and teaching a class of "foreigners," most of whom spoke English as their first language, terrified me. Therefore, I never applied for a graduate teaching assistantship during my master's program.

However, I was curious about how Canadian university instructors taught their students. It so happened that one day in early September 1998 I came across a registration form from the UTS (University Teaching Services) offering "free" sessions to faculty, sessionals, librarians, postdoctoral fellows, graduate students, and others interested in teaching. Several topics attracted my attention and I registered to attend these sessions. I continued to do so throughout my master's program.

Three years went by and I obtained my master's degree in Education. By then, I had become more familiar with the Canadian academic setting through participating in various departmental activities, both academic and social. I sat on committees and helped organizing parties. From observing several different Canadian professors teach and doing my own presentations at graduate seminars, I slowly felt more and more comfortable speaking the English language and regained my confidence as a qualified university instructor in the process. I decided to officially enroll in the UTP (University Teaching Program[2]) at UTS when I began my doctoral studies in September 2001. I attended altogether more than fifty hours of presentations, seminars, and workshops that prompted me to examine a range of theoretical teaching topics and to imagine what it might be like for me to teach in a Canadian university classroom. When the Department of Secondary Education offered me the opportunity to coteach an ESL curriculum and methods course the following September, I felt I was ready.

Narrative Moment #1: Losing My Teacher Voice

On the morning of September 5, 2002, I dressed in my teaching suit and arrived on campus at 7:00 am. I headed to my classroom to ensure that the tables and chairs were set up for group work and the overhead projector was next to a table where I placed handouts and transparencies.

When the class began at 8:00 am, my coinstructor and I each introduced ourselves. I mentioned that I would videotape my teaching as part of a UTP requirement and asked for their signatures on a consent form.

The class proceeded according to our lesson plan with each of us taking turns leading group activities and discussions, and giving mini-lectures about Lee Shulman's Pedagogical Content Knowledge (Shulman 1987) and Stephen Krashen's six hypotheses of second language acquisition (Krashen 1991).

Class ended at 10:30 am and I returned to my office to watch the video. I was shocked to find out that I had lost my teacher voice over the past four years! Compared to my coinstructor's voice, mine was too soft, lacked assertion, and force. She spoke far louder and clearer and with more confidence than I did. I also noticed that I did not provide clear instructions as to what the students needed to accomplish by the end of the class. The grouping and seating arrangements were problematic and inefficient. A table with six students was far too crowded. I looked very uncomfortable when the students were having group discussions. I did not know how to join their conversations. I just walked around, appearing rather awkward.

Thus began my journey of learning to teach in a Canadian university classroom. I still remember how anxious I felt before that first class. Even though I knew the curriculum content I was going to teach, I was very worried. I had so many questions and felt as though I had very few answers. How would I be able to help my student teachers, most of whom were native English speakers, learn to teach in a context that was foreign to me? I was not born here. I never went through the Canadian school system. I had no idea how and what a secondary English as a second language (ESL) teacher would teach to a class of students who might come from all four corners of the world and speak different first languages. I did not know what a typical school day would look like for them. How would I establish credibility that they could and would learn something useful from me? How would I be able to accommodate their learning needs? What would be my role and relationship with them? What kind of behavior is expected of students in a Canadian classroom? When the academic year comes to an end, how does a teacher categorize a good student?

Looking back, I realize that all of these questions are related to my feelings of inadequacy as a nonnative speaker teaching native-speaking undergraduate students and my not knowing the new teaching context. I was not sure what to do or where to position myself while teaching in that classroom. Although I had ten years of university teaching experience in China and knew the content I was teaching very well, I felt very uncomfortable in both my body and mind on that day. Things that I used to take for granted as a teacher in China were suddenly disrupted and I found myself struggling in a space of uncertainty and ambiguity, not knowing how to be a teacher in that Canadian classroom. Moving onto a strange "professional knowledge landscape" (Clandinin and Connelly 1995) I did not realize then that it was my "personal practical knowledge" (Connelly and Clandinin 1988) that was called into question. "Personal practical knowledge" is

> a term designed to capture the idea of experience in a way that allows us to talk about teachers as knowledgeable and knowing persons. Personal practical knowledge is in the teacher's past experience, in the teacher's present mind and body, and in the future plans and actions. Personal practical knowledge is found in the teacher's practice. It is, for any one teacher, a particular way of reconstructing the past and the intentions of the future to deal with the exigencies of a present situation. (Connelly and Clandinin 1988, p. 25)

For me, my personal practical knowledge as a university teacher was embedded in and composed through my experiences of schooling and teaching in Shanghai, China. I knew in both my body and my mind what it meant to teach in a university classroom in China and how I should do it. I also knew how students should behave in my classroom, how they should set their goals and what was important to them in their lives. I was a knower (Vinz 1997) then. I was very comfortable with my old "stories to live by" (Connelly and Clandinin 1999), stories with plotlines of teacher as a sage on a stage, of students as empty receptacles to be filled up with knowledge and of achievement as measured only by test scores. "Stories to live by" is a term that helps me to "understand how knowledge, context and identity are linked and can be understood narratively" (Connelly and Clandinin 1999, p. 4). In my case, my old stories to live by reflected how the places, times, and relationships shaped who I was as a student and who I was as a teacher and how I taught in China (Cui 2006). These stories were largely invisible to me until that first class when I experienced the discomfort in both my body and my mind on the new Canadian professional knowledge landscape. "To enter a professional knowledge landscape is to enter a place of story," write Clandinin and Huber (2005, p. 47). I did not realize then that I needed a set of new stories to live by as a university teacher in Canada and that these stories would only come with time and experiences in this new context. But I was determined to learn.

I cannot remember what else I did during the following eight weeks when I taught those twenty-two student teachers for two-and-a-half hours twice a week. I spent most of my time planning and preparing lessons with my coinstructor, teaching classes, and marking assignments. Half way through the course, I began to notice that my teacher voice was coming back, that my body was more relaxed, and that I moved with more ease in the classroom while interacting with my Canadian students. I was happy to see those changes in me. I realized then that it was possible for me to learn how to teach in a Canadian university. My hope of becoming a teacher educator in a Canadian university began to glimmer.

By April 2003, I had finished most of the requirements of the UTP and all my coursework for my doctoral program. I had become more confident in what I knew as a university instructor in Canada. More importantly, I had become a regular member at the weekly Research Issues Conversations at CRTED for two years. Every Tuesday between 12:30 and 2:00 pm, I would join this community of learners, mostly graduate students, who sat around a big table to share ideas, stories, uncertainties, and successes of their lives as teachers and/or as qualitative researchers. At first, I was a silent listener more than an active participant, not knowing how to join the conversations. Not until I felt safe and comfortable in this learning community did I begin to hear my own voice.

> As we listened to each other's stories and told our own, we learned to make sense of our teaching practices as expressions of our personal practical knowledge, the experiential knowledge that was embodied in us as persons and was enacted in our classroom practices and in our lives. It was knowing that which came out of our pasts and found expression in the present situations in which we found ourselves. (Clandinin 1993, p. 1)

I finally found myself an academic homeplace, a place where I could reclaim my teacher voice. It was there that I realized the importance of attending narratively to the stories we live and tell as teachers and teacher educators and the power of their educative potential for growth and change (Clandinin 2002, p. 18). It was there that I realized who I was/am/will be as a person was/is/will be interconnected with who I was/am/will be as a teacher and teacher educator, and that my learning and teaching experiences on both the Chinese and Canadian university education landscapes would continue to inform and shape how I would teach in a classroom here in Canada.

I decided to apply for the principle instructor position for the same ESL curriculum and methods course for the 2003–2004 academic year. When I learned that I would, indeed, be teaching the course in the fall, I began planning and preparation right away because I had so many ideas that I wanted to try out in my *own* classroom. September came really fast, so did October and November.

Narrative Moment #2: Reclaiming My Teacher Voice

On November 4, 2003, I dressed in my teaching suit and arrived at my classroom at 7:00 am for my last class of the term. As usual, I rearranged the rows of chairs into groups of four for my class activities. I then organized my teaching materials on a table and wrote my agenda on the blackboard. I waited for my class to begin at 8:00 am.

I had told my students that there would not be a final exam and that this final class would be our time to share and celebrate our learning together during the previous eight weeks. We completed the first activity and then had a discussion. I then asked them to work in groups of four to share one activity that they had each planned before class in order to get some responses from their peers. Next, I asked them to move their chairs into a semicircle. There were three students who would like to share with the class their language learning and/or teaching stories. As usual, I gave them each three minutes to talk about their experiences, what they had learned themselves about the process of teaching and learning of languages, and how that knowledge would influence their future teaching practices. Their classmates responded to their stories with questions and comments. We then played a review game on the two major frameworks that were discussed throughout the term, revisiting the concepts, terms, theories, and ideas. I was happy to notice their confidence when speaking about what they had learned in the course. I handed out my last feedback sheet and asked for their suggestions as to how I might teach this same course in the future.

I ended the class by reading a poem by Allan Glatthorn to my students:

> *What is the Teacher?*
> What is the teacher?
> A guide, not a guard.
> What is learning?
> A journey, not a destination.
> What is discovery?
> Questioning the answers, not answering the questions.
> What is the process?
> Discovering the ideas, not covering content.
> What is the goal?
> Open minds, not close issues.
> What is the test?
> Being and becoming, not remembering and reviewing.
> What is the school?
> Whatever we choose to make it.

I gave them each a copy of the poem and wished them good luck with their first teaching practicum.

Looking back, I still remember a sense of when everything began to come together for me as an educator in a Canadian university. I no longer felt flustered when I stood in front of the class. My voice was clear and I spoke with confidence. I knew where to position myself for a lecture or during discussion, and how to relate to my student teachers both in and out of the classroom. Most importantly, I knew enough then to feel that it was alright for me to "not-know" (Vinz 1997, p. 139) and to try to figure things out along the way. My hope of becoming a teacher educator in Canada began to float.

When I returned to my office after that last class, I was eager to read my students' suggestions for my future teaching. Twelve out of those seventeen students chose to write me a letter on the back of the feedback sheet. I was really pleased when several of them mentioned that they enjoyed this course because of the friendly, open, and warm environment I created through many group activities. It seemed that these activities had allowed the class to become a community of learners that cared about each other. Most of them felt ready for their first teaching practicum. They also made some good suggestions for how I might do things differently in the future. What surprised me the most, however, was several students wrote that the reason my class was their favorite during that term was that it was so different from most of the classes they had experienced at the university! Most classes were in lecture format with mid-terms and finals and my class was nothing like that! Surprised, I immediately began to wonder, why Canadian university classrooms were more "traditional" than I had imagined.

I still wonder today where my ideas of what a Canadian university classroom should look like came from. Being new on the Canadian educational landscape and assuming that Canadian university classrooms should be different from Chinese ones, I was able to teach my Canadian student teachers the way I had believed they should be taught without realizing that I was improvising along the way, "rather than pursuing a vision already defined" (Bateson 1989, p. xi). I become aware that my personal practical knowledge grows and flourishes through narratives of my teaching experiences on both the Chinese and Canadian educational landscapes. I become aware that I am in the process of composing my teaching life (Vinz 1996), nourished by the rich experiences that teaching in both China and Canada have offered me. I have come to trust that these experiences give me a teacher voice that is uniquely my own. I am no longer afraid of stepping into the often frightening spaces of uncertainty and ambiguity because I know now these can be spaces of unexpected and delightful possibilities and this gives me hope.

Further Thoughts: Moving Ahead with Hope

According to Barbara V. Nunn (2005), a philosopher, a person's hopes are "narratively consistent with his life story or personal narrative" (p. 73). For her, every human being has a set of stories, or action scripts, available to her that she has access to, can imagine, or can use to explain human action. These stories to live by are largely determined, or at least, constrained, by the society within which she is located. They "make certain futures possible, or likely, or imaginatively available, for that human being" (Nunn 2005, p. 72). As a newcomer Chinese graduate student learning to teach in a Canadian university, my personal narrative as a university teacher was disrupted when I moved from Shanghai to Edmonton ten years ago. For a while, I found myself in a story-less space of uncertainty and ambiguity, not knowing what I could or should hope for on this new landscape. As I found and created a new set of stories to live by, I was able to envision new possibilities, and my hope of becoming a university teacher in Canada was strengthened.

Looking back, there are many factors that have contributed to fostering and supporting my hope of becoming a teacher educator in Canada. Completing coursework for my graduate classes as well as the pedagogical requirement of the UTP through presentations, seminars, and workshops allowed me to imagine what it might be like for me to teach in a Canadian university classroom. Coinstructing with my Canadian-born colleagues and observing several others teaching the same course provided me with the opportunity to see the ways they relate to students, use anecdotes to explain and clarify ideas and concepts, and display confidence as qualified university instructors. These experiences helped me to imagine how I should teach, act, and respond to students in a Canadian classroom. Working alongside my UTP teaching mentor Dr. Marg Iveson, I became more aware of both my strengths and the areas for improvement as a teacher. Having two terms of teaching in a Canadian classroom provided me with opportunities to test my new teaching skills and further develop as a confident and effective teacher. Compiling my first teaching dossier helped me to reflect on my teaching philosophy, experiences, and accomplishments in order to set short-term and long-term goals so that I could continue to grow professionally. Participating in the weekly Research Issues Conversations at CRTED helped me to realize the importance of informal learning by sharing stories about my teaching practices and listening to others.

It took me three years to officially complete all of the requirements for the UTP. Now I realize how much I have struggled to learn about a Canadian way of teaching at every step of the journey. In the end, it was my passion about teaching and my hope that I would make a difference in my student teachers' lives that sustained me. As a result of this learning journey, I also realize that there is not one right way of teaching. I am becoming a better teacher educator and this gives me tremendous hope as I look forward to my new teaching position at the Faculty of Education at the University of Manitoba in Canada.

Notes

1. Every Tuesday, from 12:30 to 2:00 pm, the Centre for Research for Teacher Education and Development, where Dr. Jean Clandinin is the director, hosts ongoing weekly "research issues" conversations for graduate students and faculty to share and refine their research projects.
2. The University Teaching Program is a free and voluntary program that is directed primarily to graduate students who are teaching assistants so that they can develop an ethical, philosophical and practical basis for careers in postsecondary teaching. It is made up of three components: pedagogy, practicum, and teaching dossier. The program must be completed while the student is registered in a graduate program at the University of Alberta. For a full description of the program, please see http://www.gradstudies.ualberta.ca/utp/about.htm, retrieved May 18, 2008.

References

Bateson, M.C. 1989. *Composing a life*. New York: Plume Book.

Clandinin, D.J. 1993. Teacher education as narrative inquiry. In D.J. Clandinin, A. Davies, P. Hogan, and B. Kennard (Eds.). *Learning to teach, teaching to learn: Stories of collaboration in teacher education* (pp. 1–15). New York: Teachers College Press.

———. 2002. Teacher education as narrative inquiry. In J. Huber, D.J. Clandinin, M. Huber, K. Keats-Whelan, D. Labbe, S. Murphy, and P. Steeves. *Reshaping classroom and school contexts: learning from stories of aboriginal children and families* (pp. 18–31). Teaching and Learning Research Exchange, Project #69. Saskatoon, SK: Dr. Stirling McDowell Foundation for Research into Teaching.

Clandinin, D.J., and F.M. Connelly. 1995. *Teachers' professional knowledge landscapes*. New York: Teachers College Press.

———. 2000. *Narrative inquiry: Experience and story in qualitative research*. San Francisco: Jossey Bass.

Clandinin, D.J., and J. Huber. 2005. Interrupting school stories and stories of school: Deepening narrative understandings of school reform. *Journal of Educational Research and Development* 1(1), 43–61.

Clandinin, D.J., P. Steeves, and S. Chung. 2007. Creating narrative inquiry spaces in teacher education. In B. Johnston and K. Walls (Eds.). *Voice and vision in language teacher education: Selected papers from the Fourth International Conference on Language Teacher Education* (CARLA Working Paper Series, pp. 17–33).

Minneapolis: University of Minnesota, The Centre for Advanced Research on Language Acquisition.

Connelly, F.M., and D.J. Clandinin. 1988. *Teachers as curriculum planners: Narratives of experience.* New York: Teachers College Press.

———. 1999. Knowledge, context and identity. In F.M. Connelly and D.J. Clandinin (Eds.). *Shaping a professional identity: Stories of educational practice* (pp. 1–5). New York: Teachers College Press.

———. 2006. Narrative inquiry. In J. Green, G. Camilli, and P. Elmore (Eds.). *Handbook of complementary methods in education research* (pp. 477–87). 3rd ed., Mahwah, NJ: Lawrence Erlbaum.

Cui, H.G. 2006. *A narrative inquiry into three teachers' experiences of learning and teaching English in China.* Ph.D. diss., University of Alberta.

Johnson, K.E., and P.R. Golombek. 2002. Inquiry into experience: Teachers' personal and professional growth. In K.E. Johnson and P.R. Golombek (Eds.). *Teachers' narrative inquiry as professional development* (pp. 1–14). New York: Cambridge University Press.

Krashen, S. 1991. *Fundamentals of language education.* Torrance, CA: Laredo.

Nunn, B.V. 2005. Getting clear what hope is. In J. Eliott (Ed.). *Interdisciplinary perspectives on hope* (pp. 63–77). New York: Nova Science.

Shulman, L.S. 1987. Knowledge and teaching: Foundations of the new reform. *Harvard Educational Review* 57(1), 1–22.

Vinz, R. 1996. *Composing a teaching life.* Portsmouth, NH: Boynton/Cook.

———. 1997. Capturing a moving form: Becoming as teachers. *English Education* 29(2), 137–46.

CHAPTER 14

A Tail of Hope: Preservice Teachers' Stories of Expectation Toward the Profession

Andrea M.A. Mattos

In face of the recent Brazilian socioeconomic panorama, the teaching profession has been experiencing threats of all sources, both in relation to how the Brazilian society views teaching and how teaching professionals view themselves. Research on Language Teaching and Language Teacher Education has reflected this trend lately in papers that deal with the issues of exhaustion, sickness, and "burnout" in language teachers (Allwright 2006; Carlyle and Woods 2003; Medgyes 1994). The results of research of this kind seem to describe the contexts where teachers may experience sickness and burnout only too well, but they hardly offer alternatives. Recent research on hope in the areas of Psychology and Education, however, shows that professionals might find alternatives and reclaim emancipatory, creative ideals through their stories of hope and expectation (Miller and Larsen 2006). This chapter seeks to show how reflecting on hope may help to provide language teachers and teachers-to-be with a sense of new possibilities and expectations that enhance their empowerment toward the profession. Through the narratives of a group of undergraduate EFL[1] teachers, who participated in a course on language teacher education, this chapter tries to show how participants find hope in the profession they have chosen.

Why Hope?

Perhaps one could think that hope is not a topic for research. Indeed, Ford (2006) says that, before the early 1990s, the topic of hope was not considered a searchable term. However, according to Turner (2005, p. 509), studies on hope date back to the 1960s. For example, Crumbaugh and Maholic (1964) have described hope as "a state that gives meaning and value to life," Stotland (1969) says "hope is motivating," and Lynch (1965) states that "as we hope we are confident of and expect a good outcome." Besides, a quick look through the Hope-Lit Database (http://www.hope-lit.ualberta.ca/)—a recently launched and now the most comprehensive public database on hope in the world (Ford 2006), developed by The Hope Foundation of Alberta and the University of Alberta Faculty of Education—shows that studies on hope are being conducted in several countries worldwide. There are several reports on studies on hope from Canada, United States, the United Kingdom, Australia, Finland, France, China, Bosnia, Belgium, Mexico, Italy, Germany, and many others. Moreover, there are numerous studies on hope in such different areas as psychology, philosophy, medical care, and nursing, as well as in sociology and education. It's easy to see that interest in the study of hope has exploded the world over. In Brazil, however, studies on hope are scarce, if not altogether absent.

But why is there such an interest in studies on hope? Why is hope so *en vogue*? Why exactly should we research hope? Well, we live in dark times (Giroux 2005; West 2004). Throughout the world, societies of all kinds are facing war, terrorism, inequalities, starvation, disease, energy crisis, environmental disasters, and so many other threats to life and peace. In Brazil, we are only too familiar with the so-called urban wars, especially in Rio and Sao Paulo, but also in other cities throughout the country. We turn on the TV and we hear of economic problems, political problems, educational problems and problems of all sorts. The teaching profession, specifically, has been facing permanent threats. Teachers in general, and public school teachers in particular, including language teachers, receive very low salaries, work long hours, endure poor working conditions, and are forced to tolerate disrespect from both their students and the society as a whole. For these reasons, teachers usually have low self-esteem and a growing number of them tend to be demotivated toward the tasks involved in teaching. These problems generally form a picture of the profession that is not very encouraging for those who are preparing to become teachers.[2]

In face of the situation described above, we need to look for sources of enlightenment. As West (2004, p. 18) has put it, "we need to be very clear about the vision that lures us toward hope and the sources of that vision."

We *need* hope. The world current historical moment faces a crisis that makes hope an obvious topic for research. Let's start by learning about what exactly hope is and how the study of hope can help to empower and engage teachers and student teachers in this quest for a better future for all involved.

The Nature of Hope

According to Larsen (this volume), "hope has been described as the ability to envision a future in which one wishes to participate" (Jevne and Zingle 1994; Larsen, Li, and Mattos 2007). Research overwhelmingly indicates that hope is a vital component of learning and successful change (Cheavens, Michael, and Snyder 2005). The energy and action that accompany hope help to make better futures possible both individually and collectively. Turner (2005, p. 509) says that "hope has been studied rather extensively from philosophical, theological, psychological and sociological perspectives." Some of these studies have used Discourse Analysis as methodological tool (Eliott and Olver 2002, 2006) and, in the area of language teaching and learning, Richards and Lockhart (1994) have mentioned the role of hope and expectations as part of the learner's belief system.

According to Sillito (2005), Dufault and Martocchio (1985, p. 379) define hope as "a *multidimensional* dynamic life force characterized by a *confident* yet *uncertain* expectation of achieving a future *good* which, to the hoping person, is *realistically* possible and *personally significant*."[3] The author explains that these multiple dimensions include cognitive, relational, affective, contextual, behavioral, and temporal spheres. Although there are several other ways in which hope has been described and conceptualized, many of these conceptualizations include a notion of hope as "an expectation that what is desired is also possible" (Turner 2005, p. 509). Among the main characteristics of hope present in the literature, hope is commonly described as futuristic, motivating, self-sustaining, pervasive, and necessary to human life (Turner 2005).

Wang (2000, p. 248) says that hope is "a common human experience," universally lived. However, it is uniquely lived or experienced by each person. This means that, although it is possible to say that all humans have experienced hope in some way, this experience is always perceived and understood from a uniquely individual perspective. The author, a researcher and university professor in the area of nursing, also says that hope is "a motivator or a power source that is positively correlated with effective coping, survival, and recovery" (p. 249). Many other researchers on hope in the areas of nursing and medical care[4] have also shown that hope is positively linked to health: hope promotes healing, facilitates coping processes, and enhances the general quality of life.

According to Nekolaichuk, Jevne, and Maguire (1999), hope is action-oriented and may be linked to other positive indicators, such as self-esteem. In their research, these authors looked at hope "from the subjective perspective of how individuals give meaning to the concept" and found that hope is represented as a "holistic, interconnected configuration of different hope elements." The authors highlight the "qualitative experience of hope" (p. 602), and agree with Wang (2000) in calling attention to the unique, dynamic nature of hope as a personal experience.

In a study involving young people's perceptions of hope, Turner (2005) reveals four possible "horizons of hope." The author labels these horizons "at-one-with; a driving force; having choices; and connecting and being connected" (p. 510). The study aimed at understanding the meaning and essence of hope for the participants and was conducted through interviews that were dominated by the participants' stories of hope. The first horizon, "at-one-with," represents confidence in life and self and was experienced by the participants in the study in several ways, such as "knowing that things were right in their world" or "a sure and confident belief that life would go well" (p. 510). The second horizon, "a driving force," represented the participants' goals and dreams that formed the basis for hoping. The third horizon described hope as "having choices," which was linked to participants' "achievement of future possibilities" (p. 511). The last horizon described in the study was hope as "connecting and being connected." For the participants in the study, this horizon reveals their understanding of humans as social beings and a necessity to establish a connection with others.

Turner (2005, p. 509) emphasizes "the importance of attaining and maintaining hope" and spots a slight difference between *hoping* and *wishing*. She says that, although the literature is inconclusive to whether there is a clear-cut difference between the two phenomena, some authors have argued that "our desires are hope if they are realistic, and wishes if they are unrealistic" (p. 513). Sillito (2005) mentions that hope is also different from optimism, exactly because "hope is realistic and does not deny the circumstance." While all these studies reinforce the importance of hope in our lives, Turner (2005, p. 513) reminds us that "lack of hope for the future, or hopelessness, is a major barrier to successful adaptation." Sillito (2005) and Miller and Larsen (2006) also subscribe to this opinion.

A further possibility in the study of hope may be the issue of exploring hope both in explicit and implicit ways. The work of Miller and Larsen (2006), for example, approaches hope explicitly, that is, both researchers attempt not only to consciously use the concept of hope but also to elicit conscious use of hope-related language and ideas by their participants.

On the other hand, Li (2007, this volume), in a study that explores her own stories of hope through her life as an ESL[5] learner and teacher, approaches hope in a most implicit way.

In order to better understand the concept of hope as possibly experienced by teachers and student-teachers in Brazil, a small-scale study was designed with the broad aim of eliciting stories and reflections on hope. This study is described in the following sections.

Learning about Hope through Narratives

The study design was based on the definitions and assumptions of the works already mentioned but especially on Turner (2005), who described hope as having four possible horizons: "at-one-with; a driving force; having choices; and connecting and being connected" (p. 510).

The context of the study was a four-month course on EFL teacher education I was offering for the English major at the Faculty of Languages, Federal University of Minas Gerais. The participants were the eighteen undergraduate students enrolled in the course who explicitly consented to participate in the study.[6] Most of the participants were student teachers with no experience in teaching, but some were already experienced teachers of English at local schools and language institutes. During the course, the students were assigned readings on themes related to teacher education and held seminars and discussions on the proposed texts, as well as on related topics proposed by the students themselves. The objective of the readings, seminars, and discussions was to foster reflection and raise awareness of important themes related to the teaching profession as a whole, but especially relevant to teaching professionals in Brazil. The students were also assigned other activities, such as classroom observations and a short Action Research Project, but these activities were not directly relevant for the purposes of this study.

At the end of the course, students were supposed to take a final test. Instead of using a formal test to evaluate students' acquired "knowledge" on the topics discussed during the course, I decided that it would be more coherent with the objectives of the course to give them a final reflective activity. This activity contained reflective questions related to the course themes, including the Action Research Project, but also some questions that were meant to collect the data for this study. These questions were based on the work of Miller and Larsen (2006) and were formulated to elicit participants' narratives (Clandinin and Connelly 2000) of hope as individually experienced and perceived by each of them. The questions were also formulated with the further objective of eliciting explicit hope-related language and ideas. The questions were the following:

- What stories from your life experience give you hope for being a teacher? What experiences at the university help you have hope for being a teacher? What experiences support your hopes for teaching? What experiences threaten your hope?
- Thinking about the topics that were discussed during the semester, what do you hope that being a teacher will mean to you and to your (future) students?
- In face of all the problems that our country is going through nowadays, and also thinking about the problems specifically related to the teaching profession, how do you hope to contribute to your profession?

The participants in the study were not required to answer all the reflective questions. On the contrary, they were allowed to choose which questions they wanted to answer, but they were supposed to answer at least three of them. Many of the students in the course chose to answer the questions that were more related to the course content and to the Action Research Project. Some of them, however, were willing to reflect on hope and were able to write rich accounts of their hope experiences. After collecting their final reflective activity and upon reading their texts, I realized that what I had was a handful of stories that help us understand the meaning of hope for these teachers and teachers-to-be, and how they find hope for teaching. The stories that follow are only a small selection of the richness and hopefulness found in their accounts.

Findings

In their answers to the reflective questions listed above, the participants shared with me their hope stories. These stories generally referred to participants' life or university experiences that gave them hope for being a teacher. Some of the participants' stories yielded similar findings as the study by Turner (2005), in that the meanings of hope underlying participants' stories were *hope as a driving force* and *hope as connecting and being connected to people*. Sometimes stories of hope come intermingled with stories of hopelessness (Miller and Larsen 2006). The stories in this study are not different: some of the stories also referred to participants' life or university experiences that represented threats to their hopes. However, participants also talked about how they would turn these threats into more hope for their envisioned futures. Below, I will present some excerpts[7] from the stories these students so generously have offered me.

Hope Stories

One of the participants in the study, for instance, says she has had many experiences that support her hopes for being a teacher. The example she cites is full of emotion and passion. The excerpt below is only a small portion of this experienced teacher's story of hope and enthusiasm for her profession:

> In my life as an English teacher I have had countless experiences that have given me hope for being a teacher. When you teach beginners that do not know the verb to be and at the end of the first class you can see the glow in their eyes because they can say "My name is..." or "My teacher is..."—there is not a better sensation, and that sensation of accomplishment comes every time you see a student leave your class knowing more about the language than he did when the class began. (M.V.)

This excerpt shows how hope is explicitly approached in this participant's story. However, some of the stories told by other participants tend to approach hope in a more implicit way, as it is the case in the following example from a less experienced student teacher:

> I had never thought of teaching until I started studying English and later, [I was] invited to become a tutor at the language institute [where] I studied. This opportunity made me see that teaching was a great area in which you share experiences and deal with human beings in the most different ways. You teach, but you also learn a lot. (R.P.)

Although this participant does not mention the word hope or any other related concept, it is possible to perceive through her story that the experience she recounts is an example of something that has given her hope for becoming a teacher.

Hope as Connecting and Being Connected

Many of the stories told by the participants share a common meaning with the stories told by the participants in Turner's (2005) study. The examples below show excerpts from some of these stories, in which participants refer to the sensation of connecting and being connected to other people, in this case, their own EFL students. Moreover, they all say that what supports their hopes for teaching is exactly this connection with their students.

> I really enjoy being a teacher and I love the relationship I have with my students. They certainly contribute a lot in giving me hopes to continue with this job. (A.M.)
>
> What gives me hope to go on teaching is the relationship I have with my students, their caress[8] with me, to see them learning and to see that they are satisfied with their learning process. (L.I.)
>
> What supports my hopes for teaching is knowing that I can help someone to learn. (D.A.)

Hope as a Driving Force

A second meaning expressed in the participants' stories was *hope as a driving force*. This finding is also coherent with the findings in Turner (2005). The following excerpts are examples of the stories that show how hope may be a driving force for these participants' professional ideals:

> (...) teachers must believe that we can do something for improving the teaching profession in our country. (...)
>
> Of course there are bad things too (...). But I am hopeful: I believe that there is always something that we can do. (D.A.)
>
> It is hard to make changes in our country but we have to believe that it is possible to do something more and to turn our work and our classes more attractive and more respectable among the students. (F.O.)

In these excerpts, participants talk about the necessity of doing something to improve the educational environment in Brazil, a reality that they are all very aware of, either as students or as teachers. Although some of them do not use the word *hope* explicitly, we can perceive that there is always something that drives them in the direction of "believing that it is possible," that they need to cling to their hopes for a better future, both for themselves and for the generations to come.

Threats to Hope

Some of the stories, however, also talked about hopelessness. One of the reflective questions was, indeed, formulated explicitly with the intention of fostering reflection on experiences that would in some way threaten participants' hopes. Their stories invariably referred to the chaotic situation of the teaching profession in Brazil. Interestingly enough, although participants mentioned the poor salaries, the long working hours, and the ill-behaved nature of students in general, what seems to be most important for these teachers is the fact that the teaching profession is not getting as much respect as they think it deserves.

> Teaching is becoming something very difficult in our country. Not only because students do not want to study hard or because of behavior problems in the classroom. Politically speaking, teaching is something really hard to do nowadays. The salary is not good and teachers, most of the times, have to work long hours to make some money. Besides, we are not respected as we would like to be, especially the foreign language teachers. (A.V.)

> We live in a country which does not invest in education. Being a teacher in such a place is not easy. Teachers are not valued, but more serious yet is the fact that teachers themselves do not value their own profession and their work. (M.V.)

Threats Becoming Hope

Amazingly, the participants in this study do not show signs of despair. On the contrary, they seem to believe it is possible to do a good job as a teacher in Brazil, even in face of the dark situation they see ahead of them. Brazilians are said to be generally hopeful. Some of the stories the participants wrote reflect this belief in a better future. Moreover, these stories show that they believe the possibility of changing the current situation lies in their very hands. The excerpt below is an example of how hopeful this teacher can be, and how she turns the dark reality into a driving force that only gives her more hope for clinging on.

> The fact that we have a problematic situation in Brazil, in relation to the low value that the teaching profession receives, only makes me feel more determinate in doing my best in order to change this situation. (...) I hope always to improve my teaching and then show to the society the importance of teachers in the process of changing the precarious condition of Brazilian education. I really believe things can be changed. (C.A.)

The title of this chapter is a reference to these stories that show how participants turn threats to their hope into more hope for a better situation. It is obviously a pun with the words tail/tale, intended to refer at the same time to the participants' narratives (tales) and to their faint hopes for success in the teaching profession.

Final Remarks

As the discussion in this chapter shows, research on hope is extensive and expanding in several academic areas all around the world. In Brazil, however, this does not seem to be the case. Obviously enough, it would be difficult to draw conclusions from such an incipient field of study in our country. Some final remarks may be of relevance, though.

The study reported in this chapter shows that language teachers in Brazil are generally hopeful toward their profession, although they recognize that the teaching environment is not very promising. The participants in this study also reveal that they find it important to go on believing in their ideals. Not only do they refuse to accommodate to the situation, but, through their hope stories, they also reveal that they have faith in a better future.

Miller and Larsen (2006) state that talking about hope, that is, using the language of hope intentionally, helps to foster hope in people. Several other researchers agree that hope stories and conversations may elicit hope in difficult situations. What we need to do, then, as researchers and teacher educators, is to bring hope into our courses and classrooms and thus help to enhance hope in the lives of our preservice and in-service teachers.

Affect is certainly a concept that has made its way through into research in Applied Linguistics and Language Teaching long ago. Hope may be studied as part of this concept field. Carter and McCarthy (2004) cite research in the area of psychotherapeutic counseling to say that "creative language choices can create paradigm shifts in awareness and perception" (p. 83). Using the language of hope, talking about hope with our teachers and teachers-to-be, is one of these cases in which we can make creative language choices with the explicit objective of providing an arena for this shift in awareness and perception that the authors talk about. They also state that "creative language choices compel recognition of the social contexts of their production: principally, the construction and maintenance of interpersonal relations and social identities" (p. 66). Doing research and fostering reflection on hope may help to construct and maintain interpersonal relations between the teacher educator and the student teachers and among the student teachers themselves. Talking about hope in our contexts may help to construct a hopeful environment for ourselves and for those around us (Daloz 2000).

In closing this chapter, I want to cite one more excerpt from the participants in this research. This teacher is one of the participants who already had a lengthy experience in language teaching. Although she is aware of the many problems involved in teaching in Brazil, she is also very conscious of the challenges entailed in overcoming such circumstances. Her plea is certainly hope-inspiring.

If we have not been achieving our goals, we need to keep the seeds of hope in our minds and hearts. Things don't change so fast. It is important to struggle against injustice that is threatening our desire to teach, by giving our best to our students. Education now will guarantee the future of our country. We can't give up. (S.D.)

Notes

1. English as a Foreign Language.
2. See, for example, Gimenez (2005).
3. Italics in the original.
4. Several examples are listed in Nekolaichuk, Jevne, and Maguire (1999) and Ting (2006).
5. English as a Second Language.
6. In this chapter, participants' identities will be preserved by concealing their names. Only the first letters of their names will be mentioned.
7. The excerpts cited in this chapter were only very slightly modified in order to avoid possible misunderstandings from the part of the reader. The symbol [ ] represents comments or words added by the researcher to clarify ambiguous sentences or phrases. The symbol (...) indicates that irrelevant words or phrases were omitted. All the excerpts were originally written in English.
8. Here, the word "caress" does not mean any physical feeling, but simply that the students care for their teacher in a sincere way.

References

Allwright, D. 2006. Prioritising the human quality of life in the classroom: Is it asking too much of beginning teachers? Conference presented at I CLAFPL (Congresso Latino-Americano sobre Formação de Professores de Línguas), Universidade Federal de Santa Catarina, Florianópolis-SC, Brazil.

Carlyle, D., and P. Woods. 2003. *The emotions of teacher stress.* Stoke on Trent, UK: Trentham Books.

Carter, R., and M. McCarthy. 2004. Talking, creating: Interactional language, creativity and context. *Applied Linguistics* 25(1), 62–88.

Cheavens, J.S., S.T. Michael, and C.R. Snyder. 2005. The correlates of hope: Psychological and physiological benefits. In: J.A. Eliott (Ed.). *Interdisciplinary perspectives on hope* (pp. 119–32). Hauppauge, NY: Nova Science.

Clandinin, D.J., and F.M. Connelly. 2000. *Narrative inquiry*. San Francisco: Jossey-Bass Publishers.

Crumbaugh, J., and B. Maholic. 1964. An experiential study in existentialism: The psychometric approach to Frankl's concept of noogenic neurosis. *Journal of Clinical Psychology* 10, 200–207.

Daloz, L. 2000. Transformative learning for the common good. In J. Mezirow and Associates (Eds.). *Learning as transformation: Critical perspectives on a theory in progress* (pp. 103–23). San Francisco: Jossey-Bass.

Dufault, K., and B.C. Martocchio. 1985. Hope: Its spheres and dimensions. *Nursing Clinics of North America* 20(2), 379–91.

Eliott, J., and I. Olver. 2002. The discursive properties of "hope": A qualitative analysis of cancer patients' speech. *Qualitative Health Research* 12(2), 173–93.

———. 2006. Hope and hoping in the talk of dying cancer patients. *Social Science & Medicine* 64, 138–49.

Ford, D. 2006. World's largest hope database inspires record numbers. *Express News*, January 30. Retrieved August 13, 2007, from http://www.expressnews.ualberta.ca/

Gimenez, T. 2005. Currículo e identidade profissional nos cursos de Letras / inglês. In L. Tomitch, M.H.V. Abrahão, C. Daghlian, and D.I. Ristoff (Eds.). *A interculturalidade no ensino de inglês* (pp. 331–43). Florianopolis: UFSC/ABRAPUI.

Giroux, H.A. 2005. *Border crossings: Cultural workers and the politics of education*. 2nd ed. New York and London: Routledge.

Jevne, R.F., and H.W. Zingle. 1994. *Striving for health: Living with broken dreams*. Edmonton: University of Alberta Press.

Larsen, D., Y. Li, and A.M.A. Mattos. 2007. Believing that it's possible: The power of student and teacher narratives of hope. In International Conference, No. 1, 2007, Belo Horizonte-MG, Brazil. *Book of Abstracts*… Belo Horizonte-MG, Brazil: Associação Brasileira de Professores Universitários de Inglês.

Li, Y. 2007. Learning with hope: Stories of an EFL/ESL student in China and Canada. In International Conference, 1., 2007, Belo Horizonte-MG, Brazil. 2007 *Book of Abstracts*… Belo Horizonte-MG, Brazil: Associação Brasileira de Professores Universitários de Inglês.

Lynch, W.F. 1965. *Images of hope: Imagination as healer of the hopeless*. Notre Dame: University of Notre Dame Press.

Medgyes, P. 1994. *The non-native teacher*. Hong Kong: Macmillan.

Miller, D., and D. Larsen. 2006. Hearing it slant: Hopeful stories from teacher educators and counseling educators. Symposium presented at Narrative Matters 2006 (The Storied Nature of Human Experience: Fact & Fiction), Acadia University, Wolfville-Nova Scotia, Canada.

Nekolaichuk, C.L., R.F. Jevne, and T.O. Maguire. 1999. Structuring the meaning of hope in health and illness. *Social Science and Medicine* 48, 591–605.

Richards, J.C., and C. Lockhart. 1994. *Reflective teaching in second language classrooms*. New York: Cambridge University Press.

Stotland, E. 1969. *The psychology of hope*. San Francisco, CA: Josey Bass.

Sillito, J. 2005. Images and stories of hope: Understanding hope and transformation with adult ESL learner and teachers. In Annual National Conference, 24., 2005, London, Ontario. 2005 On-line Proceedings...London, Ontario: Canadian Association for the Study of Adult Education. Retrieved August 16, 2006, from http://www.oise.utoronto.ca/CASAE/cnf2005/2005online Proceedings/CAS2005Pro-Sillito.pdf

Ting, D.Y. 2006. Certain hope. *Patient Education and Counseling* 61, 317–18.

Turner, de S. 2005. Hope seen through the eyes of 10 Australian young people. *Journal of Advanced Nursing* 52(5), 508–17.

Wang, C.H. 2000. Developing a concept of hope from a human science perspective. *Nursing Science Quarterly* 13, 248–51.

West, C. 2004. Finding hope in dark times. *Tikkun* 19(4), 18–20.

Notes on Contributors

Robert W. Blake Jr. is both the Associate Director for the Center for Science and Mathematics Education in The Jess and Mildred Fisher College of Science and Mathematics, and the Graduate Program Director Elementary Education Master of Arts in Teaching Program (MAT) at Towson University, Towson, Maryland, where he is an Associate Professor. He has a Ph.D. in Curriculum Design from University of Illinois at Chicago, IL, and as a science educator he initiated the first science professional development school in the state of Maryland. He continues to focus his work on best practices in elementary education and the preparation of science teaching in the elementary schools.

D. Jean Clandinin is Professor and Director of the Centre for Research for Teacher Education and Development at the University of Alberta. She is a former teacher, counselor, and psychologist. She is coauthor of *Narrative Inquiry: Experience and Story in Qualitative Research* and has recently edited the *Handbook of Narrative Inquiry: Mapping a Methodology*, published in 2007 by Sage Publications.

Bob Cox is Director of Staff Development at George Brown College in Toronto where he is responsible for the training and development of all faculty, support staff, and administrative staff. He currently serves as cochair and founder of the Canadian Society for Staff and Professional Development (CSPOD), and is a member of the Board of Directors for the Institute for the Advancement of Teaching in Higher Education (IATHE).

Dawn Garbett has taught Science Education to student teachers in early childhood, primary, and secondary teacher education programs in the Faculty of Education, University of Auckland, in New Zealand. She has recently been appointed as Assistant Dean Teaching and Learning to enhance the scholarship of teaching.

Sarah Haines is Associate Professor, Biology, at Towson University, Towson, Maryland. She has a Ph.D. in Zoology, from University of Georgia, and is the Director of the Center for Science and Mathematics Education, The Jess and Mildred Fisher College of Science and Mathematics. As a science educator she has dedicated her career to engaging teachers and students in best practices in environmental education. Her commitment is reflected in her credentials: President, Maryland Association for Environmental and Outdoor Education, Certified Trainer for Project Wet, Wild, and Learning Tree, Certified TEAM DNR (Maryland Department of Natural Resources) teacher trainer.

Rena Heap lectures in Science and Mathematics Education at the University of Auckland, New Zealand. She brought over twenty years teaching experience across all levels of schooling to this appointment. Her research interests explore teachers' understanding of the nature of Science.

Neil Hooley is a Lecturer in the School of Education, Victoria University, Melbourne, Australia. He works in preservice teacher education programs for primary and secondary teachers. He is committed to reconciliation between the Indigenous and non-Indigenous peoples of Australia and sees community-university partnerships as an important step to this end.

Denise J. Larsen is Associate Professor and Coordinator of Counseling Psychology programs at the University of Alberta. She is also Director of Research at the Hope Foundation of Alberta. Her primary research focuses on how hope is effectively fostered and utilized with professional helping conversations. She is a registered psychologist and maintains a counseling practice.

Yi Li, a former university English teacher in China, is currently an Assistant Professor at the Department of Curriculum, Teaching and Learning, Faculty of Education, University of Manitoba in Canada. Her research interests include teaching English as an additional language, teacher education and development, international education, narrative inquiry, and hope.

Rob Mark has worked as a teacher, staff developer, and researcher in the field of adult literacy. He coordinates lifelong learning courses in the School of Education, Queen's University Belfast, in Northern Ireland, which includes courses for teachers of literacy and numeracy. His research interests and publications are in the broad field of lifelong learning including management of education, community and work-based learning, and adult literacy.

Andrea M.A. Mattos is an Assistant Professor at the Federal University of Minas Gerais, Brazil, where she teaches English and Applied Linguistics to graduate and undergraduate students. She has published several academic papers and book chapters both in her home country and abroad. Her main research interests are in the areas of narrative research, teacher education, and critical literacy.

Faye McCallum is Senior Lecturer at the University of South Australia with research and teaching interests in child protection, health and well-being, and at-risk learners. She has worked extensively in schools with school-based mentors and preservice teachers. Her work in higher education has comprised teaching, research, and administration duties at various levels that include graduate and undergraduate programs in the primary/middle years of schooling.

Dianne M. Miller (formerly Hallman) is Professor of Educational Foundations at the University of Saskatchewan, Canada. Her research interests are educational biography, teacher history, and poetry.

Anne Laura Forsythe Moore received her Doctorate in Philosophy from the Ontario Institute for Studies in Education of the University of Toronto in 2006. She teaches preservice candidates education and philosophy courses in Toronto, Canada. She continues to research coparticipants' "Puzzle Framed," characterized by linguistic and cultural diversity within pluralistic societies.

Alan Ovens is a Principal Lecturer in the School of Critical Studies in Education, University of Auckland, New Zealand. His interests are teacher education, particularly of students in the four-year undergraduate Bachelor of Physical Education program, and practicum. His research focuses on understanding the lived experiences of students learning to teach.

Brenton Prosser is a Lecturer (Middle Years) at the University of South Australia and a key researcher in the Centre for Studies in Literacy, Policy and Learning Cultures. His research interests include middle schooling, teachers' work, narrative inquiry, and Attention Deficit Hyperactivity Disorder. Previously, he has worked as research fellow to the Redesigning Pedagogies in the North ARC research project.

Georgia Quartaro is Dean of General Education and Access at George Brown College in Toronto, Canada. She holds a Ph.D. in Clinical Psychology from York University and has extensive teaching, administrative, and clinical experience. Her research interests include perceptions of learning and change, meaningful inclusion in education and employment for marginalized people, counselor training, college-community partnerships, qualitative research strategies, and related topics.

Maureen Ryan is a professor at the School of Education, Victoria University, Melbourne, Australia. She has had roles as Head, School of Education, Deputy Dean, Faculty of Human Development, Head, School of Health Sciences and Acting Pro Vice Chancellor (External Engagement). Her research interests include Indigenous issues, youth, education, and community.

Judy Sillito is an adult educator who works in the field of English as an Additional Language. Her graduate research focused on hope as experienced by EAL learners and teachers. Currently she is the manager of language services at Edmonton Mennonite Center for Newcomers, an immigrant serving agency in Edmonton, Alberta, Canada.

Index

Abell, S. K., 48, 62
Abrahão, C., 214
Acker, S., 167, 168, 169, 175
Allen, A. R., 95, 105
Allwright, D., 203, 213
Alsup, J., 92, 105
Alvesson, M., 68, 76
Alvis, J., 126, 132, 135
Apps, J., 174, 175
Ardizzone, T., 152, 165
Arhar, J. M., 11
Australian. Department of Education, Employment and Workplace Relations, 78, 89
Axelrod, P., 169, 175
Ayers, W. C., 48, 62

Bailey, K., 44
Baird, R., 23
Baker, J., 110, 111, 115, 118, 121
Barry, H., 31, 44
Barton, G., 122
Bateson, M. C., 20, 27, 198, 200
Bennett, W., 67, 76
Benton, P., 67, 68, 76
Benzein, E., 153, 164
Berry, A., 126, 135
Black, L. L., 152, 153, 166
Blake Jr., R. W., 4, 5, 47, 48, 62
Bleicher, R., 125, 126, 135
Bloom, L., 95, 96, 105

Boal, A., 112, 121
Boice, R., 152, 166
Boman, J., 180, 190
Boostrom, R., 164, 165
Boyer, E., 169, 175
Brassett, A., 121
Britzman, D., 96, 101, 102, 105
Brookfield, S. D., 127, 135
Bruner, J., 1, 2, 5, 9, 31, 44, 49, 62, 77, 89, 93, 105, 113, 121
Bryan, L. A., 48, 62
Buckingham, M., 139, 142, 148
Bynner, J., 120, 121

Cady, S., 148
Camilli, G., 201
Campbell, L. A., 188, 189
Cantillon, S., 110, 111, 115, 118, 121
Carlyle, D., 203, 213
Carroll, T., 91, 105
Carter, K., 32, 44
Carter, R., 212, 213
Cassidy, T., 66, 76
Cheavens, J. S., 153, 163, 165, 205, 213
Chomsky, N., 35, 44
Chung, S., 192, 200
Clandinin, D. J., 3, 4, 5, 7, 9, 11, 14, 16, 19, 21, 27, 32, 44, 47, 62, 77, 88, 89, 93, 100, 105, 151, 153, 157, 165, 180, 185, 189, 192, 195, 196, 200, 201, 207, 214

Clark, C. M., 49, 62
Clarke, D., 66, 76
Clinton, K., 95, 105
Cohen-Cruz, J., 112, 122
Collins, J. B., 141, 148
Connelly, F. M., 3, 4, 5, 7, 9, 11, 14, 15, 16, 19, 21, 27, 32, 44, 77, 88, 89, 100, 105, 151, 153, 157, 165, 180, 185, 189, 192, 195, 200, 201, 207, 214
Cooperrider, D. L., 139, 148
Cox, B., 6, 7, 137
Craig, C. J., 157, 159, 163, 164, 165, 166
Crawford, J., 67, 68, 76
Creswell, J., 180, 189
Crites, S., 151, 165
Crowther, J., 108, 121
Crumbaugh, J., 204, 214
Cui, H. G., 195, 201

Daghlian, C., 214
Daloz, L., 212, 214
Davies, A., 47, 62, 180, 190, 200
Denzin, N. K., 189
Department for Employment and Learning—DEL (Ireland), 109, 121
Department of Education & Science—DES (Ireland), 109, 121
Devane, T., 148
Dewey, J., 3, 4, 5, 9, 13, 26, 27, 77, 80, 82, 89
Dezure, D., 165
Dickinson, E., 172, 175
Dillabough, J., 169, 175
Dillabough, J.-A. *see* Dillabough, J.
Dufault, K., 153, 165, 178, 181, 182, 189, 205, 214
Dugan, K. *see* Dugan, K. B.
Dugan, K. B., 124, 132, 135
Dworkin, T., 152, 165

Eagleton, T., 81, 89
Edey, W., 154, 163, 166, 178, 180, 181, 182, 189

Egan, J., 112, 121
Eisenberg, J., 48, 62
Eliott, J. A. *see* Elliott, J.
Elliott, J., 85, 89, 165, 201, 205, 213, 214
Elmore, P., 201
Emerson, S., 164, 165
European Union. Program for Peace and Reconciliation, 110, 121
Ewing, R. A., 91, 92, 105

Farran, C. J., 153, 165, 178, 181, 182, 189
Fegan, T., 111, 121
Feldman, D. B., 153, 166
Fish, S., 125, 135
Fitzgerald, S., 121
Fivush, R., 31, 32, 44
Ford, D., 204, 214
Forsythe, A. *see* Forsythe Moore, A. L.
Forsythe Moore, A. L., 4, 11, 12, 15, 16, 17, 18, 19, 24, 25, 27
Frandsen, K., 178, 181, 189
Freeman, D., 33, 34, 36, 43, 44
Freeman, M., 93, 105
Freire, P., 108, 121
Freyd, J., 175
Friend, L., 67, 68, 76
Frye, N., 16, 27

Garbett, D., 6, 123, 128, 129, 130
Gault, U., 67, 68, 76
Gee, J. P., 93, 95, 105
George Brown College (Canada), 137, 148
Giddens, A., 65, 66, 67, 70, 71, 75, 76
Gimenez, T., 213, 214
Giroux, H. A., 204, 214
Goddard, M., 91, 105
Goddard, R., 91, 105
Golombek, P. R., 192, 201
Gravel, L., 163, 165
Green, J., 201
Greene, M., 111, 121
Guskin, S., 152, 165

Hafen, B., 178, 181, 189
Haines, S., 4, 5, 47
Halkes, R., 34, 44
Hallman, D., 175
Hamel, T., 175
Hamilton, M., 108, 121
Hamilton, M. L., 135
Hammond, C., 120, 121
Hanke, J., 163, 165
Hardy, B., 113, 122
Harris, S., 82, 89
Hart, N., 188, 189
Hattam, R., 92, 105
Haug, F., 67, 76
Havel, V., 179, 180, 189
Heap, R., 6, 12, 123, 128, 129, 130
Helme, S., 66, 76
Helms, M., 126, 132, 135
Herth, K. A., 153, 165, 178, 181, 182, 189
Hilberg, P., 163, 165
Hilsen, L., 166
Hogan, P., 47, 62, 200
Holliday, A., 31, 42, 44
Holly, M. L., 11, 27
Holman, P., 148
Hooley, N., 5, 77, 82, 85, 88, 84
Huber, J., 180, 190, 195, 200
Huber, M., 200
Hughes, P., 77, 89
Hugo, G., 91, 105
Hume, L., 84, 89
Hunt, J., 91, 105

International Adult Literacy Survey, 108
International Working Group on Indigenous Affairs—IWGIA (Australian), 79, 90
Intrator, S., 174, 175
Irving, L. M., 163, 165
Iveson, M., 199

Jago, B. J., 152, 165
Jalongo, M. R., 48, 62
James, W., 2

Jevne, R. *see* Jevne, R. F.
Jevne, R. F., 152, 165, 178, 180, 181, 182, 189, 190, 205, 206, 213, 214
Johnson, J. Q., 175
Johnson, K. E., 192, 201
Johnsrud, L. K., 152, 165
Johnston, B., 200
Jones, W. P., 152, 165

Kagan, D. M., 49, 62
Kamler, B., 95, 105
Karpiak, I. E., 152, 165
Karren, K., 178, 181, 189
Kasim, R., 120, 122
Kasten, W. C., 11
Keats-Whelan, K., 200
Kennard, B., 47, 62, 200
Kessels, J. P., 48, 62
Kippax, S., 67, 68, 76
Kirkwood-Tucker, T. F., 125, 126, 135
Kohler-Riessman, C., 50, 51, 62, 93, 95, 106
Koike, D. A., 35, 44
Kooy, M., 48, 62
Korthagen, F. A. J., 48, 62, 126, 135
Korthagen, F. J. *see* Korthagen, F. A. J.
Kramsch, C., 35, 44
Krashen, S., 194, 201
Kruegar, R., 68, 76
Kurfiss, J., 166

Labbe, D., 200
LaBoskey, V. K., 48, 62, 126, 135
Labosky, V. K. *ver* LaBoskey, V. K.
Lahman, M. K. E., 152, 153, 166
Lamb, T., 108, 111, 115, 122
Lambe, T. *see* Lamb, T.
Lamber, J., 152, 165
Larsen, D. J., 7, 151, 152, 153, 154, 163, 166, 203, 205, 206, 207, 208, 212, 214
Lasky, S., 94, 105, 106
Lawler, S., 65, 76
Leggo, C., 93, 94, 106
LeMay, L., 154, 163, 166

Leonard, T., 106
Letterman, M. *see* Letterman, M. R.
Letterman, M. R., 124, 132, 135
Li, Y., 8, 191, 205, 207, 214
Lightbown, P. M., 43, 45
Lincoln, Y. S., 189
Liskin-Gasparro, J. E., 35, 44
Lockhart, C., 205, 214
Long, M., 32, 44, 45
Loughran, J. J., 124, 126, 135
Luke, A., 9
Lunde, K. S., 165
Lynch, K., 110, 111, 115, 118, 121
Lynch, W. F., 204, 214
Lyons, N., 48, 62

McCallum, F., 5, 91
McCarthy, M., 212, 213
McGee, R., 181, 182, 190
Mack-Kirschner, A., 48, 62
McMurtry, J., 169, 175
McRobbie, C., 127, 136
Magnuson, S., 152, 153, 166
Maguire, T. O., 178, 181, 190, 206, 213, 214
Maholic, B., 204, 214
Mark, R., 6, 107, 108, 111, 115
Markula, P., 67, 76
Marsh, H. E., 138, 148
Martensen, L., 166
Martocchio, B. C., 153, 165, 178, 181, 182, 189, 205, 214
Massumi, B., 103, 104, 106
Mattos, A. M. A., 4, 8, 31, 35, 43, 45, 203, 205, 214
Medgyes, P., 35, 36, 37, 41, 42, 45, 203, 214
Meek, M., 122
Menges, R., 152, 166
Mezirow, J., 214
Michael, S. T., 153, 163, 165, 205, 213
Miller, D. M., 7, 167, 203, 206, 207, 208, 212, 214
Moon, B., 91, 106

More, A. J., 77, 89
Munby, H., 49, 62
Murphy, P., 108, 111, 115, 122
Murphy, S., 200

Nekolaichuk, C. *see* Nekolaichuk, C. L.
Nekolaichuk, C. L., 178, 180, 181, 190, 206, 213, 214
Nelson, N., 163, 165
Nespor, J., 49, 62
Noddings, N., 94, 106
Norberg, A., 153, 164
Norton, M., 111, 122
Nunan, D., 33, 44, 45
Nunn, B. V., 199, 201
Nyquist, J. D., 166

Ochberg, R. L., 152, 166
Olsen, D., 152, 165, 166
Olsen, J., 34, 44
Olson, M. R., 157, 159, 166
Olver, I., 205, 214
Onyx, J., 67, 68, 76
Organisation for Economic Co-operation and Development—OECD, 78, 90, 91, 92, 106, 108, 122
Ovens, A., 5, 65

Pajares, M. F., 49, 62
Palmer, P. J., 152, 166, 174, 175
Parnell, P., 152, 165
Parse, R. R., 153, 166
Parsons, S., 120, 121
Pavlenko, A., 32, 45
Peavy, R. V., 166
Pellegrini, R. J., 105
Peterson, P. L., 49, 62
Popkewitz, T. S., 49, 63
Popovich, J. M., 153, 165, 178, 181, 182, 189
Pratt, D. P., 141, 148
Preston, J., 121

Prosser, B., 5, 91, 93, 94, 96, 100, 106
Prusak, A., 66, 76, 93, 95, 106

Quartaro, G., 67, 137

Ramsey, G., 91, 106
Rand, K. L., 153, 166
Raudenbush, S., 120, 122
Richards, J. C., 33, 43, 44, 45, 205, 214
Richlin, L., 166
Riessman, C. K. *see* Kohler-Riessman, C.
Ristoff, D. I., 214
Rose, C., 180, 190
Rosser, K., 152, 165
Rossi, T., 66, 76
Roth, W. M., 125, 136
Russell, T., 124, 126, 135
Ryan, M., 5, 77, 88

Sarbin, T. R., 106
Saveman, B. I., 153, 164
Schon, D., 169, 175
Schubert, W. H., 48, 62
Schuller, T., 121
Schutzman, M., 112, 122
Schwab, J., 11, 27
Seldin, P., 138, 141, 148
Seligman, M. E. P., 139, 148
Sfard, A., 66, 76, 93, 95, 106
Shanahan, T., 105
Shor, I., 109, 120, 122
Shorey, H. S., 153, 166
Shulman, J. H., 85, 90
Shulman, L. S., 194, 201
Sillito, J., 7, 8, 177, 179, 190, 205, 206, 215
Skoldberg, K., 68, 76
Small, J., 68, 76
Smith, D. L., 91, 92, 105
Smith, N., 178, 181, 189
Smyth, E., 175
Snyder, C. R., 153, 163, 165, 166, 178, 182, 190, 205, 213

Social Science and Humanities Research Council of Canada—SSHRCC, 175
Soroke, B., 109, 111, 115, 122
Spada, N., 43, 45
Steeves, P., 192, 200
Stern, H. H., 35, 45
Stinson, K., 23
Stone, R., 93, 106
Stotland, E., 178, 182, 190, 204, 214
Street, B., 108, 122

Tabachnick, B. R., 49, 63
Tett, L., 108, 121
Thelen, D., 152, 165
Thompson, S., 68, 76
Thomson, P., 95, 105
Ting, D. Y., 213, 215
Tisdell, E., 111, 122
Tobin, K., 127, 136
Tomitch, L., 214
Trimmer, J., 48, 63
Trumbull, D. J., 47, 48, 49, 50, 63
Turner, de S., 204, 205, 206, 207, 208, 209, 210, 215
Turner, J. L., 152, 166

Van Der Bogert, V., 152, 166
Van Veen, K., 94, 106
Vinz, R., 195, 198, 201

Wadsworth, R., 166
Walls, K., 200
Wang, C. H., 205, 206, 215
Warlow, A., 122
Weiner, G., 168, 175
West, C., 204, 215
Westbury, I., 27
Westra, K., 181, 189
Whelan, K., 180, 190
Whitney, D., 139, 148
Wilkof, N., 27
Williams, M., 77, 89

Williams, R., 81, 90
Willis, M., 126, 132, 135
Willis, P., 106
Wisniewski, R., 152, 166
Witherell, C., 94, 106
Wittrock, M., 62
Woods, D., 34, 45
Woods, P., 203, 213

Wright, D. L., 165
Wulff, D. H., 166

Yin, R. K., 85, 90

Zahorski, K. J., 166
Zeichner, K. M., 49, 63, 126, 136
Zingle, H. W., 152, 165, 205, 214